THE SACI

This book on urban water bodies, catchment areas and drainage pattern is set against the backdrop of the unprecedented heavy rainfall that severely deluged metropolitan cities and other parts of India in recent years. The recurring natural catastrophes in water-stressed cities of India and alarming rate of diminishing water bodies, wetlands and catchment areas needs a re-visit to an entire urban water-cycle. This book, thus, discusses how the processes and implementation of colonial urban development policies and projects have radically transformed the water bodies and their catchment areas – traditional water holding systems of Varanasi city. In this imperative colonial process, through the case study of Varanasi, this book mainly engages with the reasons behind the elimination of the temple tanks and ponds after the annexation of Varanasi by the British from 1775 till 1947. The book investigates the colonial notion of 'dry city', and how this notion crafted the process of separating land and water bodies, which arguably resulted in the reclamation and draining of water bodies, and also gave rise to water pollution. Additionally, the book analyzes the elimination of water bodies and loss of catchment areas through the ongoing processes of restoring the ancient city's natural and cultural heritage.

Mahesh Gogate is an affiliated researcher at Kyoto University, Japan and studying the dynamic and fluid topography of Varanasi city, India. Before academia, he worked in software, and spent many years working as a research associate.

The Sacred Waters ‘of’ VARANASI

THE COLONIAL DRAINING AND HERITAGE ECOLOGY

MAHESH GOGATE

MANOHAR
2023

Routledge
Taylor & Francis Group
LONDON AND NEW YORK

First published 2023
by Routledge
4 Park Square, Milton Park, Abingdon, Oxon OX14 4RN

and by Routledge
605 Third Avenue, New York, NY 10158

Routledge is an imprint of the Taylor & Francis Group, an informa business

British Library Cataloguing-in-Publication Data
A catalogue record for this book is available from the British Library

ISBN: 9781032523118 (hbk)
ISBN: 9781032523132 (pbk)
ISBN: 9781003406051 (ebk)

DOI: 10.4324/9781003406051

Typeset in Adobe Garamond 11/13
by Kohli Print, Delhi 110051

MANOHAR

Dedicated to
Śri Kāsī Viśvanātha
and the peoples of Vārāṇasī

Contents

Figures and Tables

FIGURES

TABLES

Acknowledgements

Undertaking this task of writing a book has been a truly life-changing experience for me and it would not have been possible without the continuous support and guidance that I received from many people. I would first like to thank my family and friends for supporting me throughout the years. I am fortunate to receive kind support and continuous encouragement, from my Aai, Baba and my younger brother Ashutosh. My wife Mizuho and my son, Nachiketa who has been by my side throughout this research and fieldwork, and without them, I would not have had the courage to embark on this journey in the first place.

I would then like to thank my supervisor, Dr Rohan D'Souza, for his guidance during my doctoral study programme. My gratitude extends to my other advisors and dissertation committee members who extended their support and guided me through all these years. Thank you to Dr Koichi Fujita, Dr Kazuya Nakamizo and Dr Sae Nakamura for being my advisors and for your time, energy and expertise. I must also thank Professor Yasuyuki Kono and Professor Mamoru Shibayama for offering me advice and introducing new developments in GIS studies. I am also very grateful to all the teaching and research staff that have taken time to discuss and enrich my research work. I gratefully acknowledge the assistance and guidance of Professor Shinya Takeda, Professor Takuro Furusawa, Professor Noboru Ishikawa, Professor Tatsuro Fujikura, and Professor Kiwamu Yanagisawa. I'd also like to extend my gratitude to the staff of the library of Kyoto University and the staff of the office of my graduate school for their support. In Japan, I also express my sincere gratitude to Shri Swami Medhasananda, for sharing his thoughts and giving useful guidance while writing the manuscript.

This research and writing a book would not have been possible without the co-operation and support extended by the people of

Varanasi where I will continue my research. My deep appreciation goes out to Shri Deepak Agrawal, Shri Vishal Singh, Dr Rana P.B. Singh, Dr K.N. Prudhvi Raju, Dr O.P. Kejariwal, Dr Pradeep Kumar Mishra, Dr Vidula Jayaswal, Dr Akhilesh Singh Raghubanshi, Dr Keshav Mishra, Dr Anil Kumar Singh, Dr Pravin Rana, Dr Mayank Singh, Shri Anil Kinjawadekar, Shri Navneet Raman, Shri Ajay Pandey, Shri Rajesh Shah, Shri Ashok Kapoor, Shri Shashank Singh, Temsutula Imsong, Shri Ashutosh Dixit, Shri Umesh Shukla, Shri Jitendra Kumar Sahani, Shri Deepu Sahani, Shri Rakesh Singh, Shri Mintu Majhi, Shri Krishna Mohan, Dr Umesh Shukla, Shri Ashutosh Singh and Shri Arpit Sarraf. I am also very grateful to all those at the office of Varanasi Municipal Corporation and Varanasi Development Authority, the staff of Banaras Hindu University and Bharat Kala Bhavan, and the staff of Jnana-Pravaha Center for Cultural Studies and Research at Varanasi. I also express my gratitude to Ms Vijayalakshmi Balankrishnan at the Indian Heritage Centre in Singapore for kindly allowing me to refer and use the map in their collection.

I would also like to extend my sincere thanks to Dr Vinay Sahasrabuddhe, Dr Anirban Ganguly, Shri Sunil Deodhar, Shri Shobit Arya, Dr Uday Gogate, Shri Ashish Sharma, Shri Anurag Gupta, Shri B. Shyam, Shri Lalit Kumar, Dr Nitin Kumar Tripathi, Shri Vinay Joshi, Shri Shantanu Gupta, Shri Sandeep Balkrishna, Shri Mansingh Kumathekar, Rachita Srivastava, and Akio Sugimoto.

Finally, thanks are due to my publishers at Manohar Publishers & Distributors for their extensive support, guidance and patience. I am truly grateful to Shri Ramesh Jain, Shri Ajay Kumar Jain, Shri Sanjay Kala and all the team members for their guidance and support.

My research has received funding from the Japan Society for the Promotion of Science (JSPS). I express my sincere gratitude to the JSPS for awarding me the research fellowship, which has enabled me to pursue my studies and fieldwork. Additionally, I have also received the support and research grant from Inter-Graduate School Program for Sustainable Development and Survivable Societies at Kyoto University and Docomo Scholarship for International Students. I thank you all.

Kyoto, Japan MAHESH GOGATE
September 2022

Introduction

This book aims to explore the steady transformation of Varanasi city and its hinterland under the broad theme of urban ecology and heritage from being a historical catchment of the Ganga River system to become, by the early decades of the nineteenth century, a city that was merely adjoining a river. This process of transformation of the city—from catchment to waterfront, involved the systematic elimination of a vast number of water bodies that were previously scattered across the length and breadth of Varanasi. The recent large scale exercise by the union government (since 2014) of restoring temple tanks and ponds on the ground, deviates from its guidelines and do not rehabilitate the notion of the 'city as a catchment'. In effect, these ecological restoration initiatives which will be referred and described as 'heritage ecologies' confirm the notion of ecology 'in' rather than being 'of' the city.[1]

This book discusses how the process and implementation of colonial urban development policies and projects have radically transformed the urban water bodies and their catchment areas—traditional water holding systems of Varanasi. In this process through the case study of Varanasi, this book mainly engages with the reasons behind the elimination of the temple tanks and ponds after the annexation of Varanasi by the British from as early as 1775 till 1947. Moreover, the research analyses the elimination of water bodies and loss of catchment areas through the ongoing processes of restoring the city's natural and cultural heritage.

PURPOSE OF THE STUDY

This book broadly explores three eras of the transformation processes. The first era looks at the period that is broadly before the colonial annexation of Varanasi. This pre-colonial context is mostly reconstructed by studying a range of writings and pictorial maps that explain the early fluvial ecology of the city. The second era concerns the period which looks at how Varanasi was sought to be 'modernized' or to put it in other words, how Varanasi was turned into a city that met European standards and expectations. This era witnesses the rolling out of urban reforms that hinged on the Victorian discourse of sanitation, safety and modernity. The third and final era of transformation engages with the question of post-Independent development exercises that include the restoration and beautification of temple tanks. It analyses the preferences and objectives for reclaiming and restoring the previously encroached temple tanks and ponds.

After 1947 various master plans were introduced for developing Varanasi city. However, the next big set of changes were initiated in 2014 when two significant development missions were undertaken by the newly elected union government: termed the Heritage City Development and Augmentation Yojana (HRIDAY) and City Development Plan for Varanasi, 2041-Smart Cities Mission (SCM). These pioneering programmes were introduced in 2015 with the accepted premise that the water bodies in Varanasi are the 'natural and cultural heritage of the city'.[2] I argue that by markedly identifying the water bodies as 'heritage' the development exercise represents a paradigm shift from the previous restoration development programmes. Interestingly, the notion of heritage which I mark as an undercurrent in the previous development exercise distinctly emerges in the ongoing restoration exercise of temple tanks and ponds.

These three eras within the ambit of the 'in' and 'of' paradigm to demonstrate the transformation of water bodies and catchment areas of the city are analysed. This approach, it is argued, will spur a deeper evaluation of the previous efforts to eliminate urban water bodies in India. This study is also significant at the juncture when, particularly in the last two decades, Indian metro cities are frequently witnessing torrential rain over short intervals, followed by flooding and waterlogging problems. Simultaneously, the municipal corporations, especially throughout the summer season, have to struggle with inadequate water storage in the reservoirs. This study, therefore, revisits the works regarding the increase of built-up areas, impervious and sealed urban landscape, shrinking urban water bodies and stagnation in municipal sewerage systems. This study tries to locate contemporary development strategies and their implementation in the colonial urban legacies that still obsess with the urban planners and policy makers.

PERSPECTIVE: CITY AND ITS ECOLOGY

Initially, cities were often considered to be the opposite of pristine nature. Urban studies, therefore, tended to overlook the 'physical nature' within urban areas (Benton-Short and Short 2008: 4). In the book titled *Man and Nature: or, Physical Geography as Modified by Human Action* American scholar and environmentalist George P. Marsh (1864: 36) argued that man is a 'disturbing agent' and his activities destroy the 'balance of nature'. He claimed that if there was zero intervention by a human, nature could restore to its original state. In his book, Marsh repetitively discussed the 'condition of equilibrium' and insisted that irrational human actions resulted in resource scarcity. The earlier ecological studies and conservation discourse was influenced by the arguments discussed in the works of Marsh and with the 'balance of nature' concept. This paradigm excluded human actions and the ecological studies mainly concentrated at the sites which were far from the human presence (McDonnell 2011: 6). This resulted in a minimal understanding of human actions and settlements. Therefore, former definitions of ecology did not mention the human element in their research and avoided urban areas. In the late twentieth century, with the emergence of the 'non-equilibrium paradigm' human components finally got added to the ecosystem.

At the time of the beginning of ecological studies, the world population was around 1.4 billion people, and today the world population is around 7.7 billion people (*World Population Prospects* 2019, United Nations). The burden on rural hinterlands that provide food, water, raw materials, and other resources to urban cities is increasing. Ecologists now started considering the ecosystem as open, dynamic, and unpredictable. The ecological scholars must include human actions at every spatial and temporal scale to understand the changes in the ecosystem.

R.S.R. Fitter, a social scientist, wrote a book titled *London's Natural History* in 1945. This research work is arguably one of the few studies which paved the way for the evolving discourse of the ecology of urban areas. Fitter elaborated on the development of the London city and how urban expansion had made a significant impact on plants and animals. The book illustrated convincingly how the wildlife was altered and adapted to the human made structures.

In its early development of the urban ecology period two research programmes; the Chicago School of Sociology (or human ecology[3]) and the Berlin School of Urban Ecology influenced the development of urban ecology[4] research (Weiland and Richter 2012: 3-6). Researchers of these two programmes were particularly interested in studying the city, its people, its surroundings and urban nature. Urban areas are most likely to harbour biodiversity at high scales and sustain diverse habitats compared to non-urban surroundings (Gandy 2013: 1306). Therefore, the urban ecologist argues for more research of built-ecosystems by amalgamating several disciplines to explore the human dimensions in the realm of the urban ecosystem (Pataki 2015).

Around 1950, ecology 'in' the city paradigm was presented. In the process of expansion of this paradigm, the notions of ecology 'of' the city emerged in Hong Kong Human Ecology Programme[5] in 1970. In the early stages of urban ecology, scientists and researchers were more comfortable with the acquaint environmental patches that are also present in urban areas. Therefore, the frameworks and methodologies of natural sites were adopted at the beginning of the research to understand the attributes of the urban areas. The researchers focused mainly on and studied places such as wetlands, parks, streams, etc., and excluded other areas. In other words, the entire urban settings are observed as mere background and not as a dynamic part of the ecosystem. This restricted approach of examination before the 1990s was later categorized as ecology 'in' the city framework (Grove et al. 2015: 1). This paradigm engages predominantly with 'biological and bio-ecological' disciplines and focuses on green and blue patches (Pickett et al. 2016: 8).

The ecology 'of' the city is a cohesive model which explores the entire mosaic of the city by asserting the city itself as an ecosystem. Therefore, the ecology 'of' the city framework needs to integrate various disciplines and explore this complex structure. While elaborating the nature of this model and how it varies from its former model, Steward T.A. Pickett et al. (2016: 8) brings the case of 'surface water stream'. The writers emphasize that the ecology 'of' the city framework will essentially study not merely the surface water stream but also examine carefully the complex network of the water streams that possibly include trenches, ducts, stormwater drainage flow, etc.

The Hong Kong study project[6] is an appropriate example to flag here where the holistic study of the entire Hong Kong city was carried out as an urban ecosystem. In the Hong Kong study project, not only the environment but the human dimension was also carefully examined. The ecology 'of' the city framework integrates ecological, social and human components and explores the constant interaction happening among these components. Timon McPhearson et al. (2016: 207) has further extended this scope and discussed the 'technical and built components' of urban areas. To study the significant transformation of the water bodies and the catchment ecology of Varanasi, this research rehearses the ecology 'in' the city and 'of' the city framework.

INDIA AND URBANIZATION

From the 1950s onwards India witnessed a massive pace of urbanization and demographic changes. In 1951, the urban population of India was around 17 per cent, and within six decades, more than 31 per cent of the Indian population now live in urban cities.[7] Even though many developing countries have reasonably managed the complex process of urbanization, still urban areas are often perceived as dense and ungoverned populated areas, generating unrestrained poverty and eventually destroying ecology (Mohan and Dasgupta 2005: 217). Drawing parallels between cities and concentrated settlements primarily with increasing inhabitants who dissipate the ecosystem and generate waste, generally obscures the discourse of urban policy and environmental issues (Marcotullio and Solecki 2013: 12).

In the context of making sustainable cities, one can briefly revisit the discussion on what constitutes the urban city and the term 'urbanization' as well as its limitations. It is rather challenging to define what is urban, and the definition of urban varies widely across countries and continents. Nevertheless, there are some approaches like considering the administrative boundary or urban agglomeration or the built-up area. In Europe, most countries like France, the Netherlands, and Norway categorize urban areas based on the total number of residents.[8] In the case of Asia, in countries like Japan, defining urban areas is based on the size of the population living in built-up areas and engaged in an urban type of business.[9] The Census of India (2011) defines urban areas (other than all places with a municipality, corporation cantonment board, etc.) as all places having a population of 5,000 or more, a density of population of at least 400 persons per sq. km, and at least three-fourth of the male population engaged in the non-agricultural sector.

In the beginning, Urbanization got associated with an expansion of American cities and started appearing in the print media in the late nineteenth century. From the 1950s to the 1970s urban transition took place in Europe and now about three-fourth of the European population lives in urban areas. While debating the complex and dynamic process of 'urbanization' Karen C. Seto et al. (2013: 4) defines it as '(a) multidimensional process that manifests itself through rapidly changing human population and changing land cover'. This working definition of urbanization with all other components highlights human density and decreasing the natural and semi-natural areas. Furthermore, the borders of urban areas are expanding, and as per the United Nations report (UN 2016), nearly half of the world's population now lives in urban areas. Despite this rapid and complicated process of urbanization happening around the world, at present only 1 to 6 per cent of the total surface area of the world is urbanized. However, its impact on the ecological system is far-reaching and influential (Alberti et al. 2003: 1170).

Economic development, social development and environmental protection are three pillars of sustainable development and urbanization is linked with these three pillars.[10] Unplanned urbanization obstructs sustainable development because of the lack of infrastructure and effective policy implementation. Managing the urban areas and building sustainable cities are thus some of the significant challenges before the governments. There are numerous outcomes of urbanization that are happening at different scales and levels. For example, urbanization in India, by providing some modern solutions has removed the dependency of people on natural resources and raw materials such as fuelwood from immediate hinterlands (Nagendra et al. 2013: 66).

Water in urban areas is essentially used for daily activities and for disposing of human waste. The rising demand for freshwater compels metropolitan cities like Mumbai, to transport their water from far distances (Tiwari et al. 2015: 181). However, the municipal water supply in many Indian cities, like Varanasi and Chennai, is deficient and irregular (Srinivasan et al. 2010, Raju et al. 2015). Therefore, urban residents are compelled to rely on hand pumps, tubewells, or private water vendors. It is argued that the water table decreases due to the unsustainable withdrawal of groundwater and because of increasing impervious surfaces (Tiwari et al. 2015, Thoradeniya and Ranasinghe 2016). The local water resources get neglected and deteriorate due to excess dependency and usage of deep tubewells. Growing obstruction of natural drainage points and encroachments of rivers and floodplains disrupt the hydrological system.

However, both ecological and socio-economic studies have failed to assess and analyse the heterogeneous features of urban areas; relating to climate, soil and built structures and multidimensional facets of urbanization (Sukopp 1998, Alberti et al. 2003). Economic development will propel the process of urbanization further. Therefore, it should not be evaluated exclusively as a problem or solution, rather, a comprehensive study of urban areas and urbanization is required (McDonald et al. 2013). The ecology 'of' the city framework thus offers a comprehensive sphere to investigate the urban areas while integrating the ecological, social, built and technical components of the city.

RAIN, FLOOD AND DROUGHT

In recent years, a sudden surge in extreme weather events has bred vulnerability within several Indian metro cities. In 2014 many parts of Srinagar city were submerged after heavy rain.[11] In November 2015, more than 1,200 mm of precipitation (highest in the past 100 years), and subsequently releasing of water from the reservoir, flooded Chennai city. These critical incidents draw one's attention to the vulnerability of urban cities in India. Sunita Narain (2016: 5) pinpoints Chennai city to highlight the twofold challenges that Indian cities are confronted with; one is water scarcity and the other is flooding (Narain 2016: 4-5). Water scarcity and waterlogging are two opposite sides of the hydrological spectrum that continually questions the resilience and sustainability of the growing urban centres. The climate change[12] phenomenon is forecasted and debated in the context of the availability of water resources and the growing water demand. It is predicted that climate change is going to change the rainfall pattern severely and will disturb the water supply (Edmondson and Levy 2013). National Water Policy (2012) refers to climate change and its possible impact on the availability of water and stresses on the revival of water harvesting structures and water bodies.[13]

The study of the monsoon rainfall pattern of India and urbanization indicates that the occurrence of heavy precipitation events will occur mostly in the urban areas as compared to the non-urban areas (Kishtawal et al. 2009). In the last two decades, many states and particularly the mega cities like Mumbai, Chennai, Hyderabad, Bengaluru, Thiruvananthapuram, Kolkata have recurrently witnessed heavy downpours and extreme weather conditions.[14] According to the India Meteorological Department (IMD) in August 2020, India received the highest rainfall since 1976. States like Chhattisgarh, Madhya Pradesh, Rajasthan, Gujarat, Odisha, and Maharashtra experienced heavy rainfall triggering floods and waterlogging in many parts of their adjoining states also. IMD issued two press releases within six days in August 2020 cautioning 'heavy to very heavy rainfall' and its expected impact on various states of India.[15]

The heavy rainfall events, calamities due to flooding and waterlogging in many metro cities of India were extensively covered by various state and national newspapers and by digital media. Newspapers and reports discussed the details of incessant rainfall in metropolitan cities and causes and fatalities caused due to waterlogging.

CITY DRAINAGE

I mark the torrential rain and flooding of Mumbai city in 2005 that has brought back one's attention to revisit the dynamics of urbanization. On 26 July 2005, Mumbai city witnessed unprecedented torrential rainfall in a short timespan. The city witnessed the highest recorded rainfall of 944 mm in a single day.[16] This catastrophic event triggered the discussion on the encroachment of rivers and streams (Mithi, Oshiwara, Dahisar, etc.) of Mumbai city, loss of mangroves and inadequate storm water drainage system that intensified the inundation of the city. Soon after the considerable damage of life and property by flooding, the Maharashtra state government appointed a committee of experts to analyse the reasons behind the flooding and suggest 'short-term' and 'long-term' development plans to avoid and improve the civic infrastructure.

Interestingly, the committee headed by renowned hydrologist Madhav Chitale, delved into the discussion of 'urban hydrology' to assert how the alteration of the surfaces of urban areas seriously reduced water penetration and evaporation. The report mentioned 'global warming' briefly however; more attention was given to physical features of the city and the metropolitan wings and functionalities to handle and manage the risk. The report (re)emphasized that the city of Mumbai should be analysed as a 'coastal city' built mostly on the reclamation land during the British colonial rule, which was earlier an archipelago of seven islands. Indeed, excess rainfall is one of the critical components; however, it is essential to understand that in the initial phase of the Indian monsoon Mumbai does receive high rainfall.[17]

Harish Iyer (2017), in his newspaper article, discussed the heavy rainfall and waterlogging problems from 2005 to 2017 in Mumbai city. Iyer compared the 'flood' disaster in Assam and Bihar state with the 'waterlogging' problems of Mumbai city. Many parts of Bihar and Assam state were submerged when the enormous amount of excess water from rivers and lakes overflowed after the heavy downpour. However, he warned us that what happened in Mumbai city was not a 'flood'. Instead, it was a severe waterlogging problem that resulted after the cloud burst in 2005 and downpour events in short durations in 2017.[18] Iyer pointed out the failure of the effective storm drain system, polluted rivers and loss of mangroves that aggravated the situation in Mumbai. Mitashi Singh and Varnika Upmanyu (2019) in an online article, further developed Iyer's proposition and tagged the urban floods as a 'misnomer'. The authors recognized the climate change factor; but stressed on the 'water sensitive urban design' and effective urban planning and implementation to tackle the deluge calamities. Donald Watson et al. (2011) summarized succinctly why floods are undesirable by pointing towards the deficiency of one's understanding regarding the flood as a part of the natural system and intruding in flood-prone territories.

Donthi, Narasimha Reddy (2020), in an online news article, discussed a wide range of topics to understand the plausible reasons behind the flooding in the city of Hyderabad. Based on his investigation, he pointed out that regardless of the quantum of rainfall, the city was crippled with routine flooding. He further argued that land use changes and growing impervious surfaces increase the storm water and add massive pressure on the drainage. In this background, it has become necessary to revisit the intricate drainage patterns of Indian cities.

According to the Census of India 2011, more than 31 per cent of India was urbanized, and the Ministry of Urban Development has estimated that the rate of urbanization will increase to 60 per cent by 2050. The urban flooding document of 2017 prepared by the Ministry of Urban Development, pointed out the exaggerated phase of spatial expansion of urban extents that have magnified the flood risk.[19]

Previous studies (R. Chigurupati 2009, Anil K. Gupta et al. 2011, Rajit Sengupta 2016) discuss the climate change explicitly, urban heat island effect, unplanned urbanization, loss of urban wetlands and water bodies, inefficient drainage system, etc. Researchers and experts stress the importance of the preservation of open spaces, urban water bodies and the need for an effective storm drain system to mitigate the risk of waterlogging. This research, therefore, studies the 'flood' phenomenon in the context of Varanasi and how it got interpreted and shaped through colonial understanding. The draining of water bodies is conceived as a countermeasure to flooding.

WETLANDS AND URBAN WATER BODIES

A wide range of studies (Forman 2008; D'Souza et al. 2011; Nagendra 2013; Raju et al. 2015; Sharma et al. 2017) have discussed and explored the functions of urban water bodies such as providing fresh water, recharging groundwater, stabilizing soil, reducing peak flows, mitigating flood risk, microclimatic cooling of the urban area and its aesthetic features. Ecostystems of wetlands and water bodies are not only essential sites for biodiversity but also act as sponges[20] by soaking runoff water and recharging the groundwater table.

Researchers and environmental experts have argued extensively about the negligence of urban water bodies and wetlands in the context of recurring floods in urban cities. Unplanned growth of urban areas has disturbed the catchment areas and their natural drainage features severely which causes heavy runoff. Compared to the open spaces, impervious urban areas have around 85 per cent of rainfall turning into a runoff (Tjallingii 2012). Furthermore, the rise in a built-up area, decreasing pervious surfaces, paving of open space accelerate the waterlogging problems. In the last five decades, the urban wetlands of India have shrunk to around 30 per cent (Gupta and Nair 2011: 1644). The decline of wetlands, water bodies, and open spaces is not restrained to any city or region of India. More than 3,000 ha of wetlands in Hyderabad city have been extinct in the last 12 years. In Bengaluru city, more than 45 per cent of lakes became extinct between 1970 to 2010, and around 72 per cent of lakes have lost their catchment areas.

The 'Chennai Floods 2015' report discusses the various reasons for flooding and repetitively underlines the impact of rapid urban expansion, decline of water bodies and defective drainage systems (Narasimhan et al. 2016). In the report, it is argued that unplanned urban expansion and encroachment has affected and reduced the water holding capacity of these water bodies. As an example, the report states the construction of the runway of Chennai Airport's on the flood basin of the Adyar River, which hampered the natural flow of water. In December 2015, after the heavy rain, water was released from the reservoir which caused the inundation of the airport and neighbourhood areas. The report mentions the expansion of urban areas in unsustainable ways which significantly alter the hydrological cycle. The land surface of Chennai city is a flat coastal plain and to control the flood it needs an effective drainage system. In Chennai, the total number of stormwater drains are far less than the existing length of urban roads, which makes the city totally vulnerable to torrential rain (Narain 2016: 4-5). New urban settlements lack efficient and proper drainage facilities and often the polluted water flows into open storm drainage. Kapil Gupta (2005: 22) points out the sluggish response in improvising the municipal drainage and juxtaposing it with water supply development.

In India, water bodies are recorded as municipal assets in the 'Municipal Land Use' records. The urban water bodies and their proprietorship is one of the contested issues and causes of complex litigations. Multiple agencies like revenue department, urban development, municipal corporations categorically emphasize that the ownership of all forms of water bodies should remain with these agencies.[21] The reclamation and the process of restoration of the temple tanks, ponds, stepwells, etc., become complicated as in many cases these water bodies are an integral part of the temple complexes.

The sixth report of 'Standing Committee on Water Resources 2015-2016' mentioned about 1,000 urban water bodies that were identified for the repair, renovation, and restoration scheme.[22] However, the studies conducted by the Centre for Science and Environment (2016), highlights that the data of the total number of water bodies is not available and only a few municipal bodies have the statistics of water bodies. Additionally, the research work also argued that rapid urbanization and changes in land use would be sustained if the water bodies are preserved and protected ecologically.

RECONSIDERING VARANASI AND ITS ENVIRONS

Varanasi[23] is one of the few ancient cities with an enormously unique position. The city is located exceptionally in the middle of the vast and fertile plains of River Ganga. The city is positioned on the west side of River Ganga over a high natural embankment of limestone with an average height of 77 m (Raju et al. 2015; Singh et al. 2017).

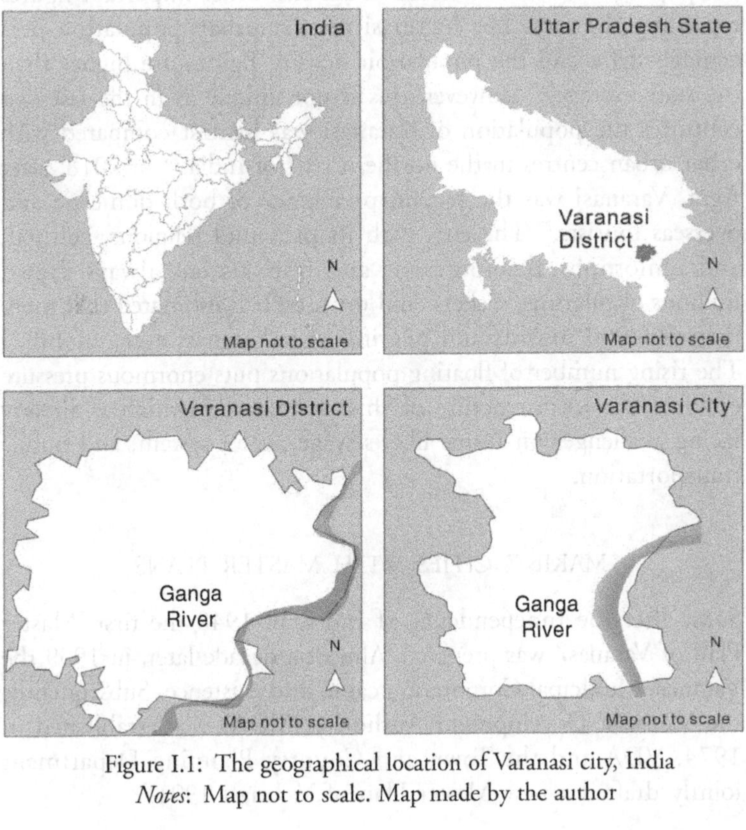

Figure I.1: The geographical location of Varanasi city, India
Notes: Map not to scale. Map made by the author

In 1940 the archaeological department carried out excavations on the north-eastern part of the city which has established the remnant of urban settlement, earthworks, and sealings. Archaeologists and scientists affirm that the river Ganga migrated in the Varanasi region around 40,000 years back (Jayaswal 2011: 15; Shukla et al. 2012: 110). Varanasi city had three considerable urban settlements: the city core area (*pucca mahal*), Rajghat and Aktha. The Ramnagar area, located in eastern side of the River Ganga with its interconnection, is considered either as a 'satellite settlement' or part of the Varanasi city (Jayaswal and Kumar 2005-6: 85).

The Varanasi Municipal Corporation (VMC) had a population of around 1.3 million people[24] (in 2013) spread across an area of 82.1 sq. km with a total of 90 wards. While discussing the size of the population, one also needs to see two other important figures regarding the city. The Varanasi district urban population percentage share and the population density figures are higher than the state average.[25] However, this is not unique as in the last two centuries the population of Varanasi was highest compared with other urban centres in the northern part of India.[26] In 2018, after Agra, Varanasi was the second preference of both domestic and overseas tourists.[27] The city, with its profound religious, cultural, and philosophical importance and festivals has always drawn millions of pilgrims, seekers, and tourists. It is estimated that more than 500,000 tourists and pilgrims visit Varanasi every month.[28] The rising number of floating populations puts enormous pressure on the civic infrastructure of this ancient city which is already facing challenges on fronts like, sewage, water system, and public transportation.

MAKING CITIES WITH MASTER PLANS

Soon after the Independence of India, in 1948 the first 'Master Plan of Varanasi' was prepared. Almost a decade later, in 1959 the Varanasi Municipal Corporation came into existence. Subsequently, the Varanasi Development Authority (VDA) was established in 1974. VDA and the Town and Country Planning Department jointly drafted a new Master Plan of Varanasi 1991.[29]

Eventually, to remove the shortcomings from the Master Plan of 1991 and for future development, a comprehensive 'Master Plan 2011' was prepared. Rana P.B. Singh (2016) in his discussion on the Master Development Plans of Varanasi, highlighted the massive expansion of the city area, especially in the last three decades that resulted in a rise in built-up areas. He further argued that more than half of the area intended for parks and open land had decreased, which would jeopardise the catchment ecology of the city (Singh 2016: 426). Singh further stated that many priority projects from the master plans were not implemented.

Varanasi was amongst the 63 cities selected for the seven-year programme titled 'Jawaharlal Nehru National Urban Renewal Mission' (JNNURM) launched in 2005 to improve the essential urban services focusing on water supply, sanitation, redevelopment of old city areas, etc.[30] The evaluation of the JNNURM scheme pointed out that the mission did not achieve significant results especially in sewage, sanitation and solid waste management of Varanasi.[31] Many renewal projects initiated under JNNURM were not accomplished[32] and projects such as sewage treatment[33] plans were later discontinued. The mission preferred the cities with large size Urban Local Bodies (ULBs) from the developed states and metro cities (Kundu 2011, 62). While analysing India's urbanization, Jagan Shah (2020) in an online article, pointed out the obsolete systems and planning in water supply and sanitation bodies that continue to carry the colonial legacy.

The guideline document of JNNURM and its language is the repetition of the earlier renewal programmes. The mission does mention the upgradation of water supply, sewage, solid waste management and preservation of water bodies. Vrinda Dar (2005: 139) discussed the notion of conservation and argued that the development agenda had outweighed the conservation practices. She stated that Varanasi has gone through a series of uncontrollable modifications and reconstruction of the built environment. She further said that the city has somehow sustained and resurrected; however, 'unsustainable and unplanned development' activities are posing new threats. The urban renewal programme continues to perceive the city in the binary of conforming and non-conforming.[34]

Thus, when dealing with pressing civic issues, the execution and implementation of programmes preferably choose to operate in the 'in' framework.

HOLISTIC MODELS AND DEVELOPMENT

In this background, in 2014, the newly-elected union government initiated various flagship programmes for inclusive, efficient, and sustainable urban habitats.[35] This book focuses on two of these pivotal programmes and their implementation in Varanasi. The first programme was the 'Heritage City Development and Augmentation Yojana' (HRIDAY). The second programme was 'Smart Cities Mission'. The Ministry of Housing and Urban Affairs was responsible for implementing these two mission programmes.

Varanasi was one of the twelve cities in India which was selected under the HRIDAY programme. In the section titled 'Need for the Scheme' it states:

> Past efforts of conserving historic and cultural resources in Indian cities and towns have often been carried out in isolation from the needs and aspirations of the local communities as well as the main urban development issues, such as local economy, urban planning, livelihoods, service delivery, and infrastructure provision in the areas. The heritage development of a city is not about (the) development and conservation of (a) few monuments, but (the) development of (the) entire city, its planning, its basic services, quality of life to its communities, its economy and livelihoods, cleanliness, security, reinvigoration of its' soul and explicit manifestation of its character.[36]

The document of Operational Guidelines 2015 emphasized on 'development of the entire city' linking it with the urban planning and execution and delivery. It also underlined the fact that the approach of the implementation was 'programme-based' and not project based. The notions thoroughly discussed in this document tells the reader that a conservation model is a holistic approach with its adaptable nature according to the needs of the city.

WATER AND HERITAGE ECOLOGIES

Urban settlements primarily grew along the rivers, freshwater lakes and near perennial water sources. These urban habitats reveal the deep cultural, spiritual, social, and ecological relationships between water bodies and civilizations. Article 48A of the Indian Constitution specifies that the state has the responsibility to protect and enhance the environment. The Indian government has introduced several policies[37] and acts to remove encroachment and for the conservation and restoration of water bodies in urban areas. The Pond Development, Protection and Conservation Authority Bill (2017) of Uttar Pradesh State provides a broad description of water bodies. The bill recognizes all water bodies (including pond, tank, lake and catchment areas) of having areas as 0.5 hectors or more and regardless of whether it holds water or not but stated as *tālāba*, tank or by any other name.[38]

The 'Draft National Water Framework Bill, 2016', states categorically that water is a 'common heritage' and innately attached with the 'people's landscape, history and culture'. The bill further adds that along with rivers, water bodies and wetlands should be considered as ecological systems and protected from disuse and degradation.[39] The City Development Plan for Varanasi 2041, restates the historical, architectural, ecological and socio-cultural dimensions of temple tanks (*kuṇḍa*) and ponds (*tālāba*) of Varanasi.[40] The temple tanks and ponds epitomize not only the architectural craftsmanship (Agarwal and Narain 1997: 116) but also the premodern technology of stormwater management. The development plan elaborates multiple functions of the temple tanks and ponds and claims that the climate change, destruction of catchment areas, blockages in inlets and outlets instigated the drying up of water tanks.[41] Thus, by bringing the water at large into the ambit of heritage and additionally embracing the ecological functions of water, the city development plan offers a new paradigm for this research.

The Archaeological Survey of India lists 18 centrally protected monuments and sites in the Varanasi district.[42] However, the list does not include a single water heritage site.[43] The natural and cultural blended landscape is an agglomeration of around 1,250 heritage sites identified by the Indian National Trust for Art and Cultural Heritage (INTACH).[44] Amongst these 1,250 heritage sites, the preparation plan of HRIDAY enumerates a total of 18 water bodies of the city.[45] The HRIDAY programme categorizes the Varanasi city area into six zones. In zone number three (Durga Kuṇḍa and Sankat Mochan Area) the programme identifies 'built and natural heritage' by mentioning three famous temple tanks; Durga Kuṇḍa, Kurukṣhetra Pokharā and Pushkar Tālāba.[46] The other total of fifteen water bodies of the city are also identified as important city landmarks.[47]

In 2016, Varanasi was identified under the flagship programme of the union government titled Smart Cities Mission. The focal point of the Smart Cities Mission is to promote a clean, sustainable, environment friendly and inclusive development.[48] The Smart City Proposal describes the ponds and lakes as sacred and treats these water storage archetypes as 'heritage'.[49] The programme discusses the rejuvenation and purification of water and real time monitoring of ten temple tanks and ponds. It allocates the fund under Corporate Social Responsibility (CSR). The project underlines the notion that the revival of temple tanks and ponds will improve ground-water recharge. Both these missions categorically highlight the heritage and ecological dimensions of the water bodies and locate them on the broader gamut of the urban environs. The HRIDAY programme and Smart Cities Mission comes as a marked departure from the 'in' framework and reinvents the conservation and development into the comprehensive 'of' model.

However, based on fieldwork and studies, one argues that the language and notions of the 'of' model reflected in the texts of these two mission programmes do not transpire in the actual processes of restoration and beautification exercise of water bodies at the ground level. Instead, the ongoing restoration and beautification exercise arguably interrupts and undermines the organic features and functions of these water heritage structures. The exercise deviates from seeing temple tanks and ponds along with its catchment areas as one broad unit. The ongoing work does not necessarily address its catchment areas which are now tough to identify and locate. The restoration works concentrate heavily on the visible components of the temple tanks and thus withdraw from the 'of' model.

VARANASI CITY AS A CATCHMENT

After descending from the Himalaya Mountain ranges, the headwaters converge with River Bhagirathi, River Janhavi and then with River Alaknanda at Devprayag. At Devprayag, the headwaters of Bhagirathi and Alaknanda acquires the name of 'Ganga'. Thus, the iconic river of India then sets for her long journey of more than 2,500 km and finally meets the sea in the Bay of Bengal. While heading towards the south-east direction, Ganga takes a sharp turn and flows northward[50] toward one of the oldest and continuously populated cities of India, Varanasi (see Figure I.2) therefore also commonly referred to as Uttara Vāhinī (flowing toward north). Varanasi is, like other major urban settlements in the Ganga basin,

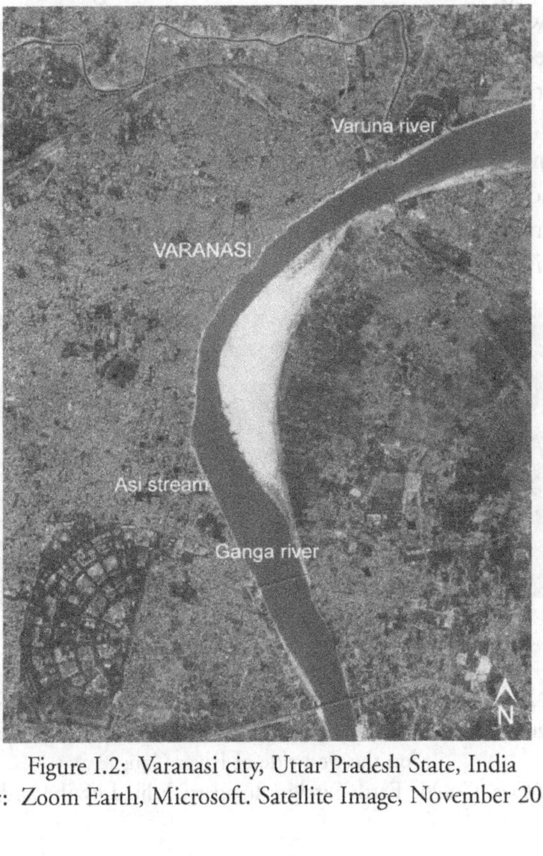

Figure I.2: Varanasi city, Uttar Pradesh State, India
Notes: Zoom Earth, Microsoft. Satellite Image, November 2018

located almost in the centre of the Ganga Valley on the west or elevated embankment on the right side of River Ganga and thus mitigating the risk of floods (Singh 1999). The flood deposits carried by Ganga on its western course eventually, developed a prominent high natural embankment (Raju and Pandey 2013: 139).

This unique geological feature of the embankment made the solid grounds and prevented the shifting nature of the settlement (Singh 1995). With its celestial features, rich legendary tales and significant geological features, the Ganga River has successfully dominated the entire floodplains and significantly influenced many ancient and sacred cities, including Varanasi that flourished on her banks. River Ganga fortified the city from invasion and offered water navigation for trade and commerce (Singh 1995).

The water of Ganga not only nurtures the civilization that has developed on its vast floodplains, but the geomorphic features of high natural levee produced several inland ponds and streams. In Varanasi, along with these inland water bodies, River Varana (Varuna) and a smaller stream (*nālā*) named Asi has a confluence with Ganga in the north and south respectively. This distinctive natural setting underpins the notion that water is a widespread phenomenon in Varanasi.

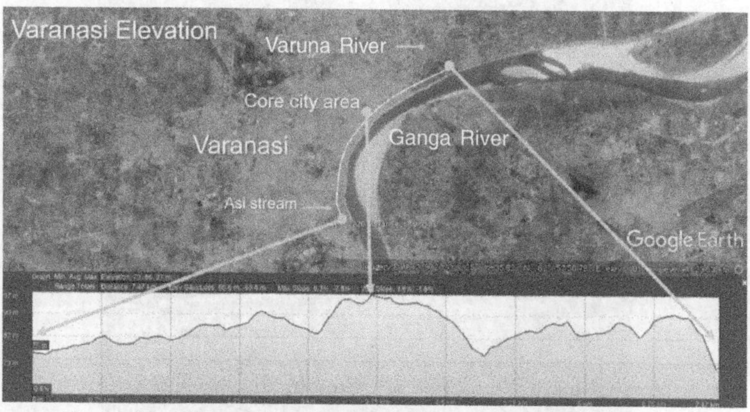

Figure I.3: Varanasi city elevation and three main embankments on the west side of River Ganga

Notes: Google Earth Satellite Imagery. Edited by the author

This natural phenomenon of high elevation appears in an allegorical way in the text Purāṇas.[51] Some of the verses describe Varanasi (Kāśī, Kāśikā) as a famous city which is on the earth but not on the ground.[52] Varanasi is heavenward and positioned on the tip of the trident (*triśūla*) of Śiva and did not submerge even during the deluge (*pralaya*).[53] These three tips of the trident are the three embankments commonly referred to as three divisions (*khaṇḍa*) of the city, namely Oṁkāra Khaṇḍa in the north, Viśveśvara Khaṇḍa in the middle and Kedāra Khaṇḍa in the south (Eck 2015).

The Rajghat plateau is on the northern side of the city close to the Varana River. The 'Chauk' area is on another highland in the core of the city. After crossing the low lying area of Godauliyā Nālā, the Kedāra Temple is located on the southern upland on the River Ganga (see Figure I.4).

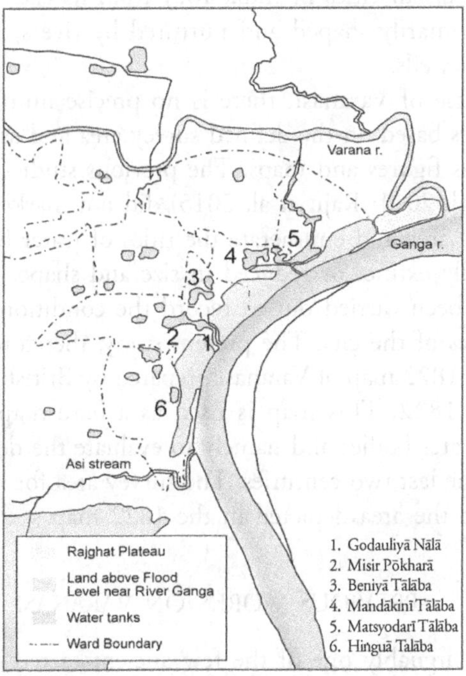

Figure I.4: Drainage pattern and geographical features of Varanasi, based on a map, in R.L. Singh, *Banaras: A Study in Urban Geography*

Notes: Tracing and edited by the author

Like other places in India, almost all forms of water reservoirs in Varanasi are closely associated with the pantheon of Hindu deities (Agarwal and Narain 1997: 281). One of the most famous wells is Jñāna Vāpī (well of wisdom), is situated in the northern direction of the present Kashi Vishwanath Temple complex. The Kashi Vishwanath Temple is one of the most famous twelve *jyotirlingas* spread across India.[54] This sacred site is the starting point for many pilgrimages that devotees make in and around the boundaries of Varanasi (Eck 2015). Various sociocultural practices, fairs and festivals are connected not merely with the water reservoirs but also with their catchment areas which suitably encapsulate the veneration of water as a cleansing agent (Aiyangar 1942).

However, particularly in the last two centuries, a surge of modern urban imagination coupled with many other colonial policies, has pushed the urban cities in India away from its catchment ecology that was primarily shaped and nurtured by rivers, its tributaries, ponds and wells.

In the case of Varanasi, there is no precise number of extinct water bodies based on the defined survey area and assessment with the previous figures and maps. The previous studies (Singh 1994; Mohan et al. 2011; Raju et al. 2015) did not markedly define the total survey area and enumerate the titles of water bodies that are either wholly extinct or reduced in size and shape. No periodical survey has been carried out to record the conditions of surviving water bodies of the city. The present study, therefore, attempts to refer to the 1822 map of Varanasi prepared by British scholar James Prinsep in 1822. This map is used as a base map; to map the surviving water bodies and namely to evaluate the decline of water bodies in the last two centuries. The survey area for this research is confined to the area depicted in the 1822 map.

PREVIOUS WORKS ON VARANASI

Varanasi is arguably one of the few ancient sacred cities in India that have continuously been visited, fondly explored, meticulously mapped, intensely written about and documented for more than last two thousand years. In this section, one starts with two of the most important scholarly works on Varanasi. The first one is Diana L. Eck's *Banaras: The City of Light* (first published in 1983) and the second is Niels Gutschow's *Benares: The Sacred Landscape of Varanasi* (2006). Whereas Eck thoroughly explored the 'sacred geography' of Varanasi, Gutschow explained the 'religious cartography' of the city. Both these voluminous scholarly works offer a painstaking analysis of the complex history of the city.

Eck has touched and expounded on many prehistoric and historical events to narrate how the city has evolved through the stories from Purāṇas, rituals and festivals, practices of worship and the development of pilgrimage routes. In the realm of the present research, her scholarship is significant mainly for two things. First, Eck has subtly explained the complex notion of 'sacred' and how geography is central to this metaphysical identity that has effectively endured for centuries. Second, her book is arguably a pioneer work on understanding the emergence of sacred geography and how this ancient city has endured its identity. Along with the sacred landscape, she has also explored how the multiple water forms have played a central role in sustaining the pure and auspi-cious identity of the city.

Gutschow's remarkable work offers an exhaustive analysis of every brick and layer that has erected the contemporary visual structural form of the city. By analysing both pilgrimage and modern scientific maps of the city, Gutschow explains how the intricate spatial organization of the city and its built forms are underpinned by the rituals, worshipping practices, fairs and festivals. Inside the larger territory of Kāśī,[55] there are a few other sacred territories. Gutschow with meticulous mapping methods and drawing on various literary sources, investigates how the city accommodates various sacred places and surpasses its peripheries and may be same can be experienced at other sacred places of India.

Both scholars have explored and discussed the water element through different lenses. While Gutschow has analysed the water bodies through the series of religious and topographic sheets and charts, Eck has dedicated an entire chapter describing the River Ganga, and the remarkable flight of masonry steps stretched on the western side of the river. However, temple tanks, ponds and other streams were discussed by both scholars as a secondary feature of the greater landscape and mainly in the context of religious importance, the evolvement of rituals and the historical connection.

Kuber Nath Sukul (1977) in his scholarly work analysed some of the earlier works that raised scepticism regarding the antiquity of Varanasi and issues related to the destruction of temples, relocation of sacred sites and temple tanks. He carefully compiled and catalogued the list of extinct and existing temples, temple tanks and several evolved pilgrimage routes.

Julia Hegewald (2002), in her comprehensive work, *Water Architecture in South Asia*, adopted the 'typological framework' to identify the five categories of water structures. She explained the history and culture of constructing these water structures, aesthetics and their diverse roles and functions in different climate regions. Interestingly, all these five typological water structures; ghāṭa[56] (flight of steps on the riverbank leading to the water), tanks, *kuṇḍa* (temple tanks) wells and ornamental pools can be found in Varanasi. Since her approach is a comprehensive analysis of water structures, she selected a few essential water structures to discuss and juxtaposed them with other analogous water structures from other regions of India and the Indian subcontinent.

Temple tanks and ponds are also studied and discussed by examining the descriptive texts of Purāṇas and several other sacred texts. Additionally, these water structures are mapped carefully analysing the available pictorial maps (Singh 1994, Gengnagel 2011). Vera Lazzaretti (2017), in her research, discussed the transition of urban waterscape of Varanasi by taking a case study of revered Jñāna Vāpī and made cursory references to a few other temple tanks and ponds. A few other studies adopted the method of a general survey of the existing water bodies and investigated the environmental change and impact of rapid urban growth (Mohan et al. 2011; Raju et al. 2015).

The gradual decline of Murshidabad,[57] intrusion and complete annexation of Benares state by the British East India Company and the 'War of 1857'[58] were some of the important events that transformed the commercial activities and strategical importance of Varanasi significantly. These events also made a lingering impact on colonial urban strategies that were unfolded afterward.

The reasons behind the decline of urban water bodies and catchment areas in Indian cities were investigated primarily and asserted chiefly in the post-Independent period of uncontrolled urban growth and expansion, coupled with the other complex outcomes of land use. The critical scholarship on the study of draining, declining and extinction of many water bodies and catchment areas of Varanasi in particular, largely overlooked the emergence of colonial discourse founded mainly on 'sanitation, safety, and modernity'.[59] The present research effort delves into the Victorian era of sanitary reforms that were initiated in Britain and later amplified in the British colonies. This overriding discourse played a crucial role in the reclamation of water tanks by delinking the network of rivers, streams, ponds and catchment areas. As a result, the functional water bodies gradually turned into 'containers' that merely hold the water. The critical examination of the British colonial policies and development projects is necessary as it exposes the favouritism of a particular water system and a process of sabotaging the other water forms and resources.

It is also essential to make a fresh analysis of the reports on the sanitary conditions of various poverty-stricken localities in Britain during the mid-nineteenth century drafted by medical doctors.[60] Janine Wilhelm (2016) argued that Chadwick's report was integrated mainly into the health priorities guidelines by the British government and marked the beginning of public health initiatives. These reports arguably set ground to build a prototype. Later, the replicas of this prototype were crafted in the British dominated colonies. This book assesses these influential reports.

The previous studies revealed that the water bodies such as ponds and human made water reservoirs in developed areas support much biodiversity and address the issues related to water management and climate change (Davies et al. 2007; Céréghino, et al. 2014; Hassall 2014). A variety of studies have discussed and explored the functions of urban water bodies such as a source of fresh water, recharging ground water, stabilizing soil, reducing peak flows, mitigating flood risk, sponge infrastructure (soaks and holds the excess amount of water) and aesthetic features (Forman 2008; D'Souza et al. 2011). However, most of these studies are framed from the compelling perspective of natural sciences, expounding upon the anthropogenic effects and individual water bodies were chosen for the in-depth inquiry. The studies on urban water bodies have paid less attention to the broader networks of general human preferences and activities, largely ignoring the roots of colonial policies. It is essential to consider various actors involving water bodies in urban areas to understand the complexity of urban water conservation and management.

It can be argued that most of the previous studies analysed and explored the riverfront heavily making cursory remarks to the inland water bodies which are embedded with the landscape and thus are more stable compared to the other physical structures. Put differently; the approach was to explore Varanasi exclusively as a 'riverfront city'. The Ganga riverfront, the series of monumental temple structures, the masonry steps (*ghāṭa*) which spread around on the western riverbank and various age-old rituals and worship practices that are also closely associated with the river and performed at the riverfront has undoubtedly captivated the attention of many European officers, authors and missionaries. During the early colonial period, the riverfront was the principal gateway[61] of the city and naturally colonial officers, and travellers wrote extensively about the River Ganga and about the city located on the splendid riverfront.

The review of colonial accounts, travelogues and journals are some of the fine examples to show how much prominence they gave towards describing the riverfront, the inner region consisting of *pukka* houses, temple architecture, military cantonment area, urban milieu, etc. In the early nineteenth century, Reginald Heber after assuming the office of the Bishop at Calcutta (now Kolkata) visited Varanasi.[62] Subsequently, many protestant and Jesuit missionaries visited and stayed for many years in Varanasi for carrying out the evangelical mission.[63] Many of these missionaries described the urban milieu, customs and showed interest in the history and geography of the second largest city in the United Provinces, during the colonial regime.

William Buyers wrote an account of his journey in his book *Recollections of Northern India* which was published in 1848. Buyers discussed various topics such as Hindu law, theology, heathen idolatry, beliefs, trade and social system which keep revolving around

Varanasi. Amongst these works, protestant missionary Reverend Matthew Atmore Sherring's work is considered as a critical scholarship on the history of Varanasi along with the discussion based on his primitive investigation of some archaeological sites in Varanasi. He was arguably the first missionary scholar who presented a detailed discussion with his prejudices and speculative claims about the antiquity of the city. Reverend Arthur Parker (1901) wrote a brief guidebook on Varanasi and lists many temples and temple tanks. Subsequently following a somewhat similar approach, Reverend Edwin Greaves described Varanasi as it was during the early twentieth century. He enumerated all major surviving temple tanks of the city and briefly mentioned the condition and festivals associated with the temple tanks.

These travelogues and accounts of Christian missionaries are essential as they show how the Europeans evaluated the ancient city and ongoing colonial progress through the spectrum of their religious dogmas and English education. Sherring and other missionaries frequently relied on the western notion and drew conversant references which corroborate their claims and denounce things that were exotic in nature. Particularly, their assessment of stagnant and filthy temple tanks and ponds influenced the English readers both in India and outside India. Their contribution and discourse made a considerable impact on the subsequent scholarship on the city.

SOURCES AND METHODOLOGY

The overarching theme of this book is to locate the premise in the paradigm of ecology 'in' and 'of' the city to explore the efficacious friction between two distinct forms of urban understandings by analysing mainly

1. literature on water bodies of Varanasi in indigenous as well as in foreign languages,
2. historical maps, satellite images, and visual material on Varanasi, and
3. data from the fieldwork undertaken by the author.

Analysis of Literature on Water Bodies

For the last many centuries, the hydrological landscape of Varanasi was characterized by its small and large water bodies, catchment areas and rivers. So far, no comprehensive study exists on the intricate relationship between the historical and structural modifications in urbanizing landscapes of the city and the significant impact on the water bodies located within. To address the reasons behind the gradual deterioration and elimination of water bodies and catchment areas of Varanasi, it is essential to understand the entire spectrum of the water cycle and how it manifests in various types of literature and other scholarly texts. Revisiting this literature and studying the description in the light of available pictorial and modern topographical maps and charts is also required. The text has been roughly categorised into indigenous and foreign texts. Indigenous texts include mainly Purāṇas, Nibandha, and other literary texts. At the same time, the foreign writings comprise of travelogues, articles, reports, gazettes produced by European travellers, officers, missionaries, and scholars.

This book attempts mainly to establish a concealed understanding of the spatial dynamics of Varanasi composed in the Purāṇas.[64] The term 'spatial text' is created to describe the qualitative geographical description. A critical review of these spatial texts with other supplementary sources is used as a foundation to build the spatio-temporal forms of Varanasi. For example, suppose a map based on the 'spatial text' can be drawn and superimposed it on the satellite imagery of Varanasi city as it is today. In that case, it is highly possible to identify and locate many surviving temple tanks and temples. This empirical exercise will further strengthen the premise of continuity of water tanks and temples that formed the sacred geography of the city. The Paurāṇika[65] texts provide an essential source of the specific site and its peculiar topographical features. This geographical description can be referred to as a fundamental premise to evaluate the subsequent changes. The allegorical description of the texts underlines the continuum between land and water without rigidly separating them from each other connotes the 'of' notion of the environment. However, these descriptive texts are stated to point of departure merely. The Paurāṇika texts are denoted as an auxiliary source to reconstruct the precolonial geography of the city.

This book attempts to make a quantitative study of the Kāśī Khaṇḍa section of the *Skanda Purāṇa.* It lists all the stated water reservoirs in Kāśī Khaṇḍa carefully. The two sections of Kāśī Khaṇḍa provides the specific geographical location of the specific temple tank concerning the other temple tanks or other water reservoirs. The orientation of the temple tank provides the sequence in the more extensive pilgrimage route and the rituals to perform at the specific site.

Surveying and Mapping Exercise of Varanasi

The larger territory of Varanasi and the eastern flood plains of River Ganga was under the control of the East India Company as early as 1775 (Blunt 1912: 21). The surveying exercise of Varanasi goes back to 1765 when a Jesuit missionary named Joseph Tiefenthaler (1710-85) visited Varanasi and did some survey of the course of the River Ganga. Later, between 1784 and 1786, Captain Brown of East India Company carried out a survey. After that in 1788, Captain Francis Wilford who was serving officer in Bengal Engineers continued the survey of the Varanasi province (Phillimore 1954). Wilford associated himself closely with Saṃskṛta (Sanskrit language) College at Varanasi. Wilford also studied Purāṇas and other literary texts with the assistance of a Hindu *paṇḍita* (scholar) to understand the Hindu cosmography and geography (Schwartzberg 1992).

Later, in 1822, British scholar James Prinsep published his map of Varanasi titled 'The City of Bunarus, surveyed by James Prinsep' which was acclaimed as the first 'accurate' map of the city. The earlier eighteenth century, topographical maps of Varanasi are not accessible. That makes the 1822 map significant to study and explore in this research project. The James Prinsep map of 1822 provides detailed information about the city and its water bodies. Therefore, a copy of the 1822 map is used as a 'base map' for georeferencing purposes.

Vera S. Candiani (2014: 30) in her work on the draining of colonial Mexico City, examined this exercise of separation because

Europeans perceived the land and water as crucial 'inputs' in the production of commodities for the trade off. Drawing on this, one can argue that, Prinsep's map does something more than this, which underpins the subsequent understanding of the city and its development strategies. The map of 1822 distinguishes water from land (Da Cunha 2019: 279). This segregation exercise withdraws from the 'of' notion and initiates the 'in' perception.

Furthermore, this book analyses two pictorial maps carefully. The first pictorial map dated eighteenth century was titled 'Map of Varanasi, Shiva's Sacred City'. The second pictorial map also dated between 1750 to 1850 was titled 'Map of Benares'. Both these indigenous pictorial maps, along with numerous temples and sacred sites depict distinct types of water structures and subtly reveal the geomorphological features of the city.

The above discussed two distinctly cartographical practices are categorized in the western scientific and accurate maps and indigenous pictorial maps or charts. Both these distinctly cartographical practices juxtapose the fundamental perception to observe the constantly shifting territories of land and water. The detailed examination of two unique representation styles offers how the fluid characteristics of the landscape were captured by the contemporary mapmakers.

Fieldwork Data

The fieldwork and physical mapping of all the existing water bodies were carried out from February 2016 to November 2019, based entirely on the James Prinsep map of 1822. The assessment of all the water bodies (excluding the River Ganga, streams, and wells) depicted on the James Prinsep map of 1822 were made by direct observation, interviews and discussions and listing all the existing and extinct water bodies. The fieldwork and research activities were conducted mainly in Varanasi, Delhi, and Mumbai city, India.

In Varanasi, all the surviving and extinct temple tanks and ponds of the city depicted on the map of 1822 were mapped. The other temple tanks and water pools which were not mapped by James Prinsep in his 1822 map were also visited and mapped during the fieldwork. Discussions and interviews with residents of Varanasi city, pilgrims, devotees and with the Varanasi Municipal Office and Varanasi Development Office personnel were conducted from February 2016 to November 2019.

CONCLUSION

This book delves into two successive periods and investigates the imperatives that shaped the water bodies and catchment ecology of the city. Before delving into the colonial period, it is essential to find out how the city was evolved and attained its unique significance that evokes contestation and evaluation of the fluvial ecology of the city and its environs. Chapter 1 explores the spatial descriptive texts such as Purāṇa-s and Nibandha-s to (re)construct the landscape of the city and how the free flowing water sustained the sacredness of the city. The chapter discusses the terms: sacred, *tīrtha* and *kṣetra* and how they are pivotal to discuss the continuum between water and land. The chapter demonstrates the pervasive feature of water and offers a glimpse of the pre-modern setting of the city. Chapter 2 examines how the celestial qualities described in literary works have shaped and produced the earthly forms of sacred places. The chapter also analyses how the spatial descriptive texts had subtly influenced the process of producing pictorial maps of Varanasi. The chapter marks the beginning of bringing the fluid geography of Varanasi and coagulating it through the surveys and map-making processes. Chapter 3 moves directly into the early period of colonial annexation and analyses the extempore urban sanitation reforms and marks the beginning of drying up of the significant temple tanks. The chapter, in many ways, is the central section of the entire book. The chapter revisits the Victorian period and cities struggling with filth, pollution and epidemics, spread through the 'toxic smell'. The chapter traces the sanitary and other urban reforms of nine-teenth-century England and their calibrated introduction in colonial cities like Varanasi. The chapter also traces the trajectory of how the Hindu pilgrimage sites were labelled as the epicentre of deadly diseases like cholera. Chapter 4 investigates the legacy of colonial planning reflected in the post-Independent urban development agendas and plans in India. By revisiting the descriptive texts, pictorial, and topographical maps and with the satellite imageries, the chapter maps how some of the temple tanks were restored in their new forms. By doing so, it tries to analyse the process of restoration and demonstrate the gulf between the text and the actual restoration exercise. The chapter argues about the continuity of the 'in' model even though the heritage document and mandate discuss and underpins the 'of' notion.

NOTES

Names of Indican cities and their English spellings were often changed during imperial rule. However, this research refers to Varanasi (including city and former state) instead of Kashi and Banaras.

The Sanskrit and Hindi words (including the temple tanks and ponds) are rendered as per the Sanskrit transliteration (excluding the nouns), and the English 's' is used as a plural marker after, e.g. Purāṇa-s.

1. Grimm et al. (2000: 573)categorises two different types of urban research namely, ecology 'in' cities and ecology 'of' cities.
2. In 1972, the General Conference of UNESCO at Paris highlighted the rising threats to the cultural and natural heritage sites in their 1972 World Heritage Convention document. The 1972 Convention identified the architectural monuments, buildings, and structures formed by both nature and men as cultural heritage. The natural heritage encompassed geological and physiographical creations and habitats of endangered animals and plants.
3. Herbert Sukopp (2002: 383) discussed urban ecology studies are mainly carried out by social researchers who conduct the social investigation and are referred to as human ecologists.
4. Initially, urban ecology was regarded as natural science (Sukopp 1998) and later developed as a distinct discipline in the early 1970s which studies living organisms and their complex relationship with the urban environment (Sukopp 2002).
5. The 'Hong Kong Human Ecology Programme' adopted the paradigm of ecology 'of' the city by integrating biological, physical as well as cultural aspects of the entire city. This unconventional approach separates it from the conventional territory of the ecology 'in' the city paradigm and highlights the serious need for an effective framework for the study of urban settlements

and reflects on the consideration of intangible aspects which are although challenging to analyze and measure in the conventional way.

6. Stephen Boyden et al. 1981, *The Ecology of a City and Its People: The Case of Hong Kong.*
7. Census of India 2011, Government of India.
8. France—Communes, consisting of an agglomeration of more than 2,000 residents living in adjacent houses. Netherlands—Urban areas with a population of 2,000 and more residents. Norway—Areas with 200 or more residents. For information on the definitions of urban, see the *Demographic Yearbook 2005.*
9. *Demographic Yearbook 2005.* The United Nations Statistics Division.
10. *World Urbanization Prospects: The 2014 Revision,* United Nations.
11. The book titled *Why Urban India Floods* has analysed the 2014 floods in Jammu & Kashmir (now union territory). The floods as not new to new to this union territory. The studies pointed out the blockages in flood channels of major rivers and the diminishing of wetlands.
12. The World Bank in its published research (2013) on climate change and its impact on India mentioned the changing rainfall patterns, predicted the drought, coastal flooding and water stress, https://www.worldbank.org/en/news/feature/2013/06/19/india-climate-change-impacts (accessed on 10 September 2020).
13. National Water Policy, 2012, Ministry of Jal Shakti Department of Water Resources, River Development and Ganga Rejuvenation, Government of India.
14. Some of the major flood calamities since 2005. Mumbai (Maharashtra): July 2005; Uttarakhand: June 2013, Jammu & Kashmir: September 2014; Bengaluru (Karnataka): September 2015: Gujarat: June-July 2015; Chennai: November 2015; Assam: July-August 2016; Hyderabad (Telangana): September 2016; Bengaluru (Karnataka): August 2017; Kerala: August 2018; Mumbai (Maharashtra): August 2020.
15. Intense rainy spell over central parts of the country during next 5 days 19 August 2020, Press Release, India Meteorological Department and heavy rainfall over east and north-west India during 25 to 28 August, 2020, 25 August 2020; Press Release, India Meteorological Department, https://mausam.imd.gov.in (accessed on 21 September 2020).
16. Fact Finding Committee on Mumbai Floods, Final Report, vol. I, March 2006, p. 13.
17. Ibid., p. 34, 38.
18. In 2015 Maharashtra state received a deficient rainfall of around 1823.2 mm. After the normal and above normal rainfall (2894.5 mm) in 2016,

this year in 2017 first two months of monsoon; June and July, many parts of the state recorded deficient rainfall. However, many parts of India and metro cities, including Mumbai, Bengaluru received excess rainfall from August till late October 2017. It is argued that the withdrawal of monsoon was prolonged and added more rain.

19. Urban Flooding: Standard Operating Procedure, 2017.
20. Richard T.T. Forman (2014) discusses the analogy of sponges with the wetlands and argues that wetlands can soak the excess water only if the soil is not saturated.
21. Churning Still Water, 2012, Centre for Science and Environment, New Delhi.
22. The report also pointed out that some of the states have not maintained these records.
23. In scriptures and various works of literature, the city also referred to as Ānandakānana (forest of bliss), Avimukta (the place never forsaken by Shiva), Rudravāsa (the dwelling place of Śiva), and Mahāśmaśāna (the great cremation ground). These different names epitomize the topographical and other salient features of the sacred region and were common in a certain period. One ephemeral renaming exercise of the city took place when Varanasi came under Muslim domination. Aurangzeb, in his regime, tried to change the name of Varanasi to Muhammadabad, but the name never lasted for long. The present name of the city 'Varanasi' was officially declared in 1956 by the Government of India. Varanasi city is the administrative headquarter of Varanasi district.
24. CRISIL Risk and Infrastructure Solutions Limited, 2015, City Development Plan for Varanasi, 2041, Final City Development Plan, p. 43.
25. Census of India 2011. Uttar Pradesh, *District Census Handbook*, Varanasi, p. 14.
26. In the nineteenth century, the city had more than 300 people per sq. km. (Robert G. Varady 1989, *Land Use and Environmental Change in the Gangetic Plain: Nineteenth-Century Human Activity in the Banaras Region*, p. 235). The various census figures from 1801 to 1941 are compiled to demonstrate the variations in the initial phase of the census exercise by various colonial officers. Refer to the Appendix section.
27. District-wise Domestic and Foreign Tourist Visits and Ranking in Uttar Pradesh in the year 2018, http://www.uptourism.gov.in/pages/top/about-up-tourism/year-wise-tourist-statistics
28. City Development Plan for Varanasi, 2041, Final City Development Plan, 2015, Ministry of Urban Development. Government of India, p. 34.
29. Town and Country Planning Department, Uttar Pradesh, India.

30. Guidelines for Jawaharlal Nehru National Urban Renewal Mission (JNNURM).
31. Report No. 15 of 2012-13—Performance Audit of Jawaharlal Nehru National Urban Renewal Mission, Ministry of Housing and Urban Poverty Alleviation.
32. Rahul Chandran, 3 June 2009, JNNURM Projects Lag Over Land, Utilities and Personnel Problem.
33. Subhash Pathak, 1 October 2019, Failed Drainage System Adds to Patna's Plight as Many Areas Remain Flooded.
34. Guidelines for Jawaharlal Nehru National Urban Renewal Mission, p. 5.
35. Heritage City Development and Augmentation Yojana, 2016, p. 18.
36. Operational Guidelines for HRIDAY (Heritage City Development & Augmentation Yojana), p. 1.
37. The Guidelines for National Lake Conservation Plan (2008), Advisory on Conservation and Restoration of Water Bodies, 2013 and Standing Committee Report on Water Resources (2016) are a few examples of the government's initiatives to build the water policy framework.
38. The Uttar Pradesh Pond Development, Protection and Conservation Authority Bill, 2017 (Third Draft). Minor Irrigation Department. Government of Uttar Pradesh.
39. Draft National Water Framework Bill, 2016.
40. City Development Plan for Varanasi, 2041, Final City Development Plan, p. 176. The Heritage City Development and Augmentation Yojana, 2016, includes open spaces in the category of water bodies which collectively form the natural heritage, p. 13.
41. City Development Plan for Varanasi, 2041, Final City Development Plan, 2015, pp. 176-7.
42. Monuments and Sites in Uttar Pradesh, Archaeological Survey of India, http://www.asipatnacircle.gov.in/uttarpradesh.html (accessed on 22 October 2020).
43. In 2009, Heritage Development Plan was introduced for the revitalisation of heritage properties of Varanasi. The Heritage Development Plan, along with other heritage sites and pilgrimage routes, included and highlighted the conservation and improvements of temple tanks and ponds.
44. Heritage City Development and Augmentation Yojana, 2016, p. 13.
45. Ibid., pp. 70, 180.
46. Ibid., p. 180.
47. Ibid., p. 70.
48. Smart Cities: Mission Statement and Guidelines, 2015.
49. Smart City Proposal Varanasi, 2016.

50. River Ganga, including Varanasi, at more than six places flows towards northward.
51. Purāṇa-s is often called 'old' and associated with myths and legends. However, many scholars have argued that this is an oversimplification of the term. Purāṇa is regarded as one of the important branches of Hindu literature and one of the ways to describe this vast literature as an 'encyclopaedia' containing stories from the ancient past. According to Ramaswami Sastry (1937: 169-71) 'Purāṇa' means that which exists from the ancient period. Sastry refers to *Chāndogya Upaniṣad* which mentions both 'Itihās' and 'Purāṇa'. Sastry further explains that the main function of Purāṇas is to explain and augment the Vedaḥ. (J.L. Shastri, 1985) defines Purāṇas, as a class of literature which represents different phases and aspects of life lived by people of diverse ages. He explains that it is a repository of ancient traditions and introduces various other subjects and notions which do not fit into the standard definitions. G.V. Devasthali (1954: pp. 291-324) explains Purāṇas as an 'old narrative' and refers to the classical definition where five elements, i.e. (1) creation of the universe, (2) recreation after the dissolution, (3) genealogy, (4) the reign of Manu as primal ancestor, and (5) history of solar and lunar dynasties. Devasthali elucidates that the Purāṇas however may not necessarily include all the above five elements. A more detailed description of Purāṇas is in the following chapter.
52. See also Eck (2015).
53. Kāśī Khaṇḍa 1.2; 5.20; 44.30 narrates the tale of Varanasi placed high on his trident by Śiva. Tagare (1996) in his voluminous work of translating the *Skanda Purāṇa* mentions in the footnotes the Rājaghāṭa plateau—site of the ancient settlement of the city, which was on higher grounds and rarely flooded.
54. The Kashi Vishwanath is worshipped in the ultimate form of 'Śiva Liṅga' which is also referred to as Viśveśvara, meaning 'Lord of the universe'.
55. Varanasi was the capital of 'Kāśī' kingdom, which was a bigger territory.
56. For a detailed discussion on the *ghāṭa*, see Julia A.B. Hegewald, 2002, 'Water Architecture in South Asia: A Study of Types, Developments and Meanings', pp. 41-72.
57. Murshidabad was the capital of Bengal *subā* (region) and one of the prominent cities for trade and commerce, during the eighteenth century.
58. This book uses the contested term 'War of 1857' which is also deliberately projected as a 'sepoy mutiny' or 'great rebellion of 1857' by colonial rulers and authors. This term has been discussed widely, and according to many historians it was the 'first independence of war'.
59. Veena Talwar Oldenburg (1984), discussed three authoritarian policies

based on safety, sanitation and loyalty that governed the policies and legislation to transform the Lucknow city, after the 'War of 1857'. Janine Wilhelm (2016) referred to safety, sanitation and loyalty to study the pollution of the River Ganga during the British regime. She underlined the serious threat posed by the 'War of 1857' and changes in the sanitation policies. She analysed the sewerage system projects of Varanasi and Kanpur city, by tracing its roots in Britain.

60. Edwin Chadwick's report (including multiple reports from other medical doctors) on the sanitary condition of the labouring population of Great Britain was prepared in 1842 and submitted to the government. The report discussed the conditions of open cesspools and sewers, filthy ponds, narrow lanes and marshy grounds of many urban areas. The report recommended improvements such as forming open and airy public squares and building gardens.
61. One of the illustrative examples is from the travel accounts of Richard Temple. Temple, like his European predecessors, approached Varanasi sailing down the River Ganga. In his brief discussion on Varanasi, he introduced and praised the city as the 'finest river frontage in India'. Richard Temple, 1898, *A Bird's-Eye View of Picturesque India.*
62. During the visit to Varanasi in 1824, Bishop Reginald Heber consecrated the St. Mary's Church located in the Cantonment area of Varanasi.
63. William Smith was the first missionary who started missionary work in Varanasi in 1816 (James Kennedy 1884: 21).
64. The English translation of the Kāśī Khaṇḍa has been the primary source for the study and for making the catalogues of temple tanks and ponds of Varanasi. However, the Sanskrit-Hindi translation text is additionally referred to verify the titles and other related information.
65. Belonging to the Purāṇas or derived from them. Apte, 1957-9, *The Practical Sanskrit-English Dictionary.*

CHAPTER 1

City of Waters

This chapter delves into the Purāṇas and other literature to situate the city in the period prior to the colonial period. It draws primarily upon the spatial texts and the qualitative interpretation of Varanasi. The descriptive texts of Purāṇas Bhuvanakośa sections, Nibandhas[1] and various other pilgrimage digests render the spatiotemporal information to reconstruct the geography of the particular region. The Paurāṇika texts contain spatiotemporal information about the settings of sanctuaries, rivers, streams, temple tanks, ponds, etc. Additionally, the pilgrimage digests composed in different periods not only mention the contemporary names and rituals to perform on the banks of sacred temple tanks but also provide the trajectory of transformation happening within and around these water sites. The brief glimpses of its sacred landscape dominated by water are essential to comprehend the waterscape of the city.

The text of Purāṇas and other works of literature composed on Varanasi unfolds the phenomenon of elevated land and a moderate slope encircled by streams, water bodies with different prototypes.[2] The texts mainly shed light on the landscape of the ancient and pre-colonial period of Varanasi. The spatial texts and allegorical terminology[3] recreate the landscape of the city where there are no rigid boundaries that defines 'wet' and 'dry'.

In the opening section, two important terms are briefley discucssed: sacredness in the context of Varanasi and its waterscape and, the theory of *tīrtha* and *kṣetra*. This exercise primarily tries to answer the question of what is 'sacred' about Varanasi and highlights the geographical notions that corroborate the foundation of the city. The discussion is set on the premise that the descriptive text in Purāṇas and other spatial texts, records subtly how the city is sanctified by the presence of extensive water streaming in various forms. Primarily, the abundant free flowing water sustained the notion of a sacred site and shed light on the shifting nature of the topography. The qualitative assessment of these texts does not necessarily draw the boundaries and demarcate the land and water. Instead, it illustrates the fluidity where water rises seasonally and recedes to rise again. The spatial texts show how the overflowing water of rivers and streams flows into the seasonal channels and confluence with the temple tanks and ponds scatter in different directions. The excess water is not perceived as an undesirable effect but as the natural cycle. Therefore, it may be argued that spatial texts encapsulate the notion of 'waters of' Varanasi. The various forms of water are described in the spatial texts. The total number of water tanks from all one hundred chapters of the Kāśī Khaṇḍa texts are also enumerated. These numerous temple tanks and water pools appear later on the visual mapping of the city. In the last part of this chapter, two pictorial maps that are based on the readings of the Kāśī Khaṇḍa, are studied.

VARANASI A SACRED TĪRTHA

In this book, the term 'sacred' is used explicitly in the context of a sanctified place, a pilgrimage place, or a place near a sacred river or a water reservoir. Instead of words such as, holy, divine and religious, the term sacred is used exclusively as an epithet to develop the unique relationship of certain places closely connected with the water reservoirs and augmented with the prehistoric events, stories and tales that revere the celestial qualities of the place.

Jane Hubert (1994: 10) in her paper, described multiple indigenous sacred sites from different continents of the world to perceive the term 'sacred'. 'What does the word sacred mean, anyway?', she asked, 'Even if we can define it in our language, to what extent is the word an adequate translation of the word or words that denote unfamiliar concepts in other cultures and religions? All concepts expressed in a specific language are necessarily limited by that language. . .' (Hubert 1994: 10). Further, she added that the term 'sacred' not only had various connotations but what it accepted and omitted also varied. In this book, the term 'sacred' is referred to in a confined way to underline certain unique places and their association with clean flowing water.

The International Union for Conservation of Nature and Natural Resources (IUCN) and UNESCO guideline document explored the 'spiritual' dimensions and their significance to the people while defining the 'sacred site' (Wild and McLeod 2008). The 2008 guideline document also proposed the following definition of 'sacred natural site': 'Areas of land or water having special spiritual significance to peoples and communities'.

The discussion on the sacred natural site touches both tangible and intangible features. It is an extensive agglomeration of spiritual notions, the abode of deities and sages, places of sanctuary for meditation and places for physical well-being. These sites also accommodate the mundane activities and temporal dimensions which are often conceived as contrary to each other. It is argued that the sacredness attached to the natural sites safeguard them from human intervention. However, it is also a fact that human intervention also enhances biological diversity.

Baidyanath Saraswati (1984: xvii) in his introductory part of the book argued that the 'French sociology of religion' scholarship of explicitly classifying all religious beliefs into sacred and profane failed to appreciate the Indian thought[4] of 'continuum between sacred and secular'. He further elucidated that, certain places, including natural features such as rivers, hills, etc., are sanctified and emerge as *tīrtha*. Saraswati claimed that pilgrimage places were fountainheads in the expansion of Indian towns in pre-historical times. Furthermore, the sacred and profane notions were cultivated together at these pilgrimage places.

To understand what is sacred in the context of Varanasi, the term *tīrtha* is useful.[5] The concept of *tīrtha* is deeply anchored in Hindu philosophy and has spatio temporal dimensions.[6] *Tīrtha* has several meanings: a passage, road; a crossing in a river, ford; a place of pilgrimage near the bank of a sacred river, sacred, saviour, etc.[7] Pandurang Kane (1953: 554, 582) mentioned the frequent references of the word *tīrtha* in *Ṛgveda*, other Vedic literature and Purāṇas. He described *tīrtha* as a place with a unique natural landscape adjacent to the water reservoirs that innately magnified the goodness. Knut Jacobsen (2013: 4) while discussing the concept of a *tīrtha*, underlined the impressive natural surroundings as an essential element of Hindu pilgrimage place. He stated that sanctity is intrinsic to the place. Richard Lannoy, in his essay on Varanasi (2001-2: 377), while expounding on the notion of *tīrtha*, discussed the intrinsic relationship between nature and culture. He highlights the exceptional features of nature and the quest of a human being to regard it as an intrinsic part of divinity.

Tīrtha is intrinsically connected with the water either in the form of sipping water before the rituals or in the form of a channel to cross. 'In the wider Hindu tradition', wrote Eck (1981: 323) and mentioned hilltops, confluences of rivers, ponds, forests to define *tīrtha* as, 'these places, particularly those associated with waters, are often called *tīrthas*, and pilgrimage to these *tīrtha* is one of the oldest and still one of the most prominent features of Indian religious life'.

The word *tīrtha* is essentially a comprehensive notion that comprises not of only of a sip of pure water but all rivers, hills, groves, lakes and wells as well. If the name of any certain *tīrtha* is unknown, then it called as Viṣṇu Tīrtha (Kantawala 1995: 26). This dynamic approach sustains and restores the *tīrthas* and allows them to accommodate different forms. The sanctity of a *tīrtha* remains eternal despite the external intrusions and destruction of the sacred physical forms. To substantiate his claim, Saraswati (1984: 68) cited the demolition of temples in Varanasi by the invaders despite which it did not disrupt the sanctity of the sacred territory.

The word sacred is implicitly associated with the purifying element of flowing water. Julia Hegewald (2002: 14-15) stated that rivers brought the celestial water that nurtured all living things, from heaven to earth and were worshipped as goddesses or gods. Amita Sinha (2011: 73) claimed that it is the 'sacred landscape' that predominantly uphold the mythical narrative compared to the verbal and textual accounts.

Regarding the sacred places, Eck (2015) used the phrase 'sacred geography' to refer to the vast landscape of the Indian subcontinent. She clarified that for the word 'sacred' there was no corresponding word in Indian languages that would precisely encapsulate the Western notion associated with the word. There are few words in Indian languages such as *pāvana* (purifying, holy), *pavitra* (holy, sanctified), *puṇya* (pure, meritorious, virtuous, righteous), etc., that can connote the expression of sacred.

According to Sukul (1977: 32), the earliest reference of Varanasi as a *tīrtha* is in the *Mahābhārata*. Vana Parva (Āraṇya Parva) section of the *Mahābhārata* has one chapter titled *tīrtha yātrā parva* devoted to the pilgrimage to sacred places (Sukthankar 1942). The chapter offers an extensive list of sacred places all over India and describes the rituals and benefits of visiting these places. The long itinerary to sacred places includes Varanasi and mentions worshipping Śiva (Vṛṣabhadhvaja) (see Figure 1.3) and a ritual bath in a water pool named as 'Kapilahrada' (82, 69)[8] (see Figures 1.1, 1.2).

The Kāśī Khaṇḍa, discusses the term *tīrtha* and presents multiple descriptions of *tīrtha*. One verse says that a *liṅga* is a *tīrtha* and a water tank is also called *tīrtha* because of its association with an idol (of deity) (*SP* KKh 97.5). In one of the chapters of Kāśī Khaṇḍa

Figure 1.1: Kupuldhara Tulao, Benares, James Prinsep, 1834, Plate 2 from the third part of Benares Illustrated. The British Library Board, London
Notes: The Kāpiladhāra Temple Tank, (Kapilahrada Tīrtha)

Figure 1.2. Kāpiladhāra Temple Tank (Photo by author, February, 2017)

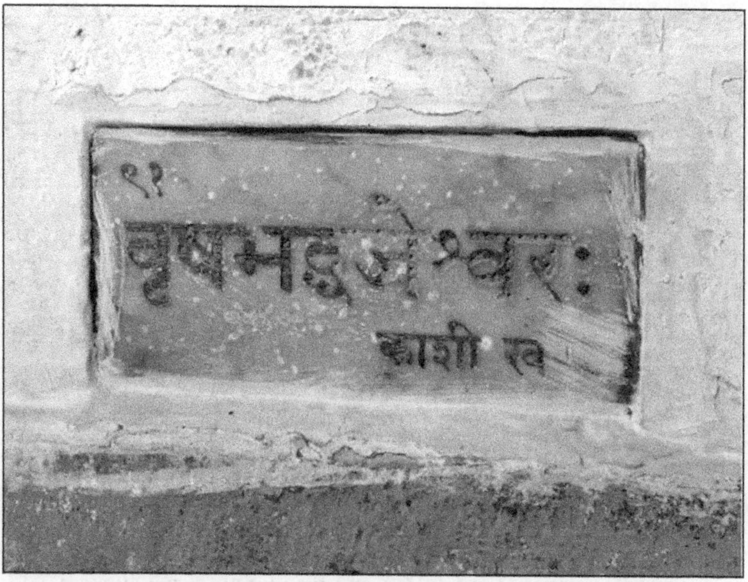

Figure 1.3: Vṛṣabhadhvajeśvara: Kāśī Khaṇḍa (Photo by author, February 2017)

sage Agastya[9] expounded on the notion of *tīrtha*. In the beginning, he described several sacred places and added that there were incorporeal *tīrthas* also. These *tīrthas* were, truth, knowledge, purity of the mind, etc. (*SP* KKh 6.29-32). He advised that ablution was not merely bathing in the water but disciplining the mind (*SP* KKh 6.33-5) (Tagare 1996).

Water is an essential element[10] of *tīrthas* that sustains the sacredness. Most sacred pilgrimage places are on the banks of a river, near a lake, or adjacent to a shoreline (Sinha 2011: 37).

TEMPLE TANKS AND PONDS

This book uses the term 'temple tanks', and it denotes *kuṇḍa, pokharā, bāvalī*, water tank, water pool, etc. The temple tanks[11] are in the proximity of the temple or part of the temple complex. Varāhamihira's (fifth century) most notable works *Bṛhat Saṁhitā* enumerates twenty types of temples and their description (Satri and Bhat 1947: 491, 496). The fifty-sixth *adhyāya* mentions that the temple site should be near the water tank and a garden. The typical shape of the temple tank is either square or rectangular with natural inlets and outlets. All four sides have masonry steps of sandstone with some design elements or clay baked bricks leading to the water. These steps are chipped in with the ramps, and the sides of the water tanks may or may not be surrounded by walls (Hegewald 2002). Many temple tanks in Varanasi have a small well in the centre, which plays a role in recharging the groundwater.

In Varanasi, one can see various architectural forms and patterns of temple tanks. Lolārka Kuṇḍa (see Figures 1.4 and 1.5) near Asi Ghāṭa has a deep tank surrounded by steep stairs leading to the water. The round well is surrounded by cylinder-shaped walls is open at the top. There is one small platform added to the series of steps. While descending the stairs, there are a few small sculptures engraved on the stones of the surrounding walls.

Among these sacred tanks two tanks named, Cakrapuṣkariṇī Kuṇḍa on Maṇikarṇikā Ghāṭa and Gaurī Kuṇḍa on Kedāra Ghāṭa are adjacent to the River Ganga. Many temple tanks also had

Figure 1.4: Lolārka Kuṇḍa
(Photo by the author, February 2017)

Figure 1.5: Lolārka Kuṇḍa
(Photo by the author, February 2017)

akhāḍās nearby. However, today few water tanks such as Īśvaragaṅgī Kuṇḍa, Karṇaghaṇṭā, Rāma Kuṇḍa (see Figure 1.6), etc., have active *akhāḍās* (see Figure 1.7). Regular members practising at *akhāḍās* recount the dynamic relationship of the *akhāḍā* with neighbouring temples and water tanks.[12]

Figure 1.6: Rāma Kuṇḍa (Photo by the author, August 2016)

Figure 1.7: Svargīya Guru Śrī Pannālāla Akhāḍā near Rāma Kuṇḍa (Photo by the author, August 2016)

The daily members and mainly older men narrated the stories of bathing in adjacent temple tanks or ponds and equated it with taking a plunge in the River Ganga. They often said that it is the water of River Gaṅgā that comes into these several temple tanks and ponds and thus taking a dip into these water tanks is also sacred.[13]

PURĀṆAS AND NIBANDHAS

The total number of Purāṇas are recognized as eighteen[14] and they contain lucid descriptions of mountains, rivers, sacred and pilgrimage places across India (Aiyangar 1942) which are regarded as an essential source in the study of historical geography (Singh 2008).[15] In particular, *Skanda Purāṇa*,[16] *Padma Purāṇa* and *Bhaviṣya Purāṇa* offer the encyclopaedic description of numerous sacred places (Pusalker 1955). Among these three Purāṇas, geography covered and discussed in Bhuvanakośa[17] section of *Skanda Purāṇa* arguably outranks the other Purāṇas (Awasthi 1965: 4). All eighteen Purāṇas have a collection of stories about Varanasi, narrating its origin and celestial aspects. Purāṇas discuss the etymology of Varanasi based on the geographical features and narrates the stories of the city. The texts offer a comprehensive description of Varanasi and its geographical features composed in the rich metaphorical phraseology. This research effort explores mainly the Kāśī Khaṇḍa, one of the seven chapters of *Skanda Purāṇa*.[18]

Brahmavaivartta Purāṇa (composed in the last millennia) has an entirely separate section of Sthāna Māhātmya[19] dedicated to Varanasi, titled as Kāśī Rahasya which was probably composed in the fifteenth century (Sukul 1977). There are various claims about the period of composition of the Purāṇas. The *Skanda Purāṇa's* oldest manuscript was dated around the sixth century CE. Baldeva Upadhyaya (1978), in his work on Purāṇa-s, suggests that the composition of *Skanda Purāṇa* took place from the beginning of seventh to the end of the ninth century. *Skanda Purāṇa* comprises of seven chapters (*khaṇḍa*),[20] and among the seven chapters, chapter four (Sthāna Māhātmya) is divided into two sections which are titled Kāśī Khaṇḍa. Sukul (1977) suggests that the period of composition of the present Kāśī Khaṇḍa occurred somewhere between the beginning or mid of the fourteenth century.

The present composition of the Kāśī Khaṇḍa is supposed to have (re)started in the late twelfth century and continued for the next two centuries.[21] Rangaswami K.V. Aiyangar (1942: lxxiv), while discussing the prominence of Viśveśvara mentioned the Telugu translation of Kāśī Khaṇḍa by Śrīnātha in the mid-fifteenth century and pointed out that the Telugu translation of Kāśī Khaṇḍa underlined the wide acceptability and eminence of the text.

Kāśī Khaṇḍa is one of the voluminous works of 'Sthāna Māhātmya' literature and a valuable source to understand the different layers and composition of an ancient city like Varanasi. *Sthāna* means a place and *māhātmya* is a part of literature that narrates the greatness and virtues of divinities and temples associated with that sacred natural place. The qualitative description of Varanasi in the chapters of Kāśī Khaṇḍa, subtly interwove the celestial and earthly features of the city which was continuously shaped and transformed by its unique location on high grounds, encircling rivers and stream and inland water pools. The texts described the spatial features and location of temples corresponding to other important sacred sites and water pools in proximity. The text carefully constructs a map for the reader to navigate and do circumambulation around the specific temples. The scrupulous attention to the text of Paurāṇika literature can throw light on the history of medieval Varanasi (Smith 2014: 23). Several chapters of Kāśī Khaṇḍa comes under the subject of Bhuvanakośa[22] and described the regional geography of Varanasi.

Here it is essential to bring the Nibandhas composed mainly in the last millennia to trace and substantiate the composition period of Kāśī Khaṇḍa The Nibandhas also denote the beginning of a profound transformation of the literature on *tīrtha*.

NIBANDHA OF LAKṢMĪDHARA BHAṬṬA

Amongst the surviving Nibandhas, *Kṛtyakalpataru* is considered as one of the authoritative and comprehensive Nibandha composed at the beginning of the twelfth century by Lakṣmīdhara Bhaṭṭa

(Sukul 1977: 233). Lakṣmīdhara was the royal counsellor (*rāja-guru*) in the court of King Govindacandra (1114-54) of the Gahadavala dynasty. Aiyangar (1942: vii) praised the eighth book titled 'Tīrthavivecanakāṇḍa' as a pioneer work on the theory of *tīrtha yātrā* (pilgrimage to sacred places). This work is a comprehensive discussion of sacred sites and sanctuaries. He adds that the composer infused a philosophy into the 'dry discussion of ritual and gradations of benefit' and this scholarly work should be evaluated not merely as an itinerary to sacred places or an ordinary manual (Aiyangar 1942).

Lakṣmīdhara devoted a substantial portion to the 'survey of the sanctuaries' of Varanasi (Bisschop 2013: 52). There are many plausible reasons behind Lakṣmīdhara placing Varanasi at the top in his composition work on sacred places. When the present digest was composed at the beginning of the twelfth century, Varanasi had already attained the reputation of one of the most revered and sacred places. Although the capital of the Gahadavala dynasty was Kannauj (Kānyakubja), Varanasi was indeed the capital in all senses (Aiyangar 1942).

According to Aiyangar (1942: lxx, lxxi), Lakṣmīdhara relied mainly on *Liṅga Purāṇa* to make a list of 340 temples.[23] Here Sukul (1977: 234) made an important observation that Lakṣmīdhara made a significant contribution by referring to *Liṅga Purāṇa* and safeguarding the records of the old version of *Liṅga Purāṇa*. The comparison of *Tīrtha Vivēcana Kāṇḍa Nibandha* with the Kāśī Khaṇḍa texts made it the oldest text that offered the spatial description of temples and other sacred water structures. Lakṣmīdhara's treatise was an authoritative source for the composers of Kāśī Khaṇḍa to rebuild the sacred landscape after the destruction. The re-structure of the text suggests that it took place probably after the twelfth century (Bakker 2009: 72).

The *Tīrtha Vivecana Kāṇḍa* composed by Lakṣmīdhara Bhaṭṭa was one of the earlier treatises from the early twelfth century followed by Kāśī Khaṇḍa a (twelfth century-fifteenth century), *Tīrthacintāmaṇi* by Vācaspati Miśra (fifteenth century),[24] *Tristhalīsetu* by Nārāyaṇa Bhaṭṭa (sixteenth century) and *Tīrthaprakāśa* by Mitra

Miśra (seventeenth century). By revisiting these texts composed in different periods one can see how the natural and cultural landscape of the city was transformed.

GEOGRAPHY IN TĪRTHA VIVĒCANA KĀṆḌA

Lakṣmīdhara made a significant contribution by mentioning various sacred sites, temples, rivers, and water pools (see Appendix 1). Lakṣmīdhara quoted from various Purāṇas while composing the text of *Varanasi-Māhātmya.* He quotes *Matsya Purāṇa* and described the Asi River as a dry river (*śuṣka nadī*) (Aiyangar 1942: 39). Lakṣmīdhara describes Matsyodarī (tīrtha) as a river flowing between rivulet Varana and a smaller stream Asi. Aiyangar (1942: lxxviii) offered a parallel argument of the Matsyodarī stream and its confluence with River Ganga in the rainy season, based on the commentary of Lakṣmīdhara. A detailed discussion on the geography of Matsyodarī and the riddle regarding its form is discussed in the next part of this chapter.

JANAPADAS

Pāṇini recorded the names of the villages, towns, *janapadas* and the mountains, forests, lakes, rivers and oceans spread across the larger territory of the Indian sub-continent. Pāṇini one of the most eminent grammarians elucidated the etymology of the name of the places. The text of *Aṣṭādhyāyī* describes the geographical information of ancient India and the role of social and historical life in establishing the names of places (Agrawala 1953: 34-36). Pāṇini has used the term *janapada* more as a 'state and cultural unity' (Agrawala 1953: 48) and less as geography.[25] According to the composers of Kāśikā[26] *janapadas* are fundamentally a group of villages and towns. Since cultural cohesion was the underpinning of political boundaries of *janapadas* they used to change. According to Agrawala (1953: 60), Pāṇini did not attribute a legend of monarchy to Kāśī *janapada* probably because Kāśī became part of Magadha. Pāṇini mentions Varanasi as the capital of Kāśī.

KṢETRA

Varanasi is significantly defined and discussed through the prism of geography. Kāśī Khaṇḍa mentions various names of the city and describes the etymology of the word.[27] Kāśī, Varanasi, Avimukta and Antargṛha are four names which often appear in Kāśī Khaṇḍa and have well-articulated boundaries. These names follow a descending order and embody unique features with well-articulated territories. Kāśī is the greater territory and was one of the prominent *janapadas.* Kāśī and another neighbouring *janapada* Kośala used to refer together as Kāśī Kośala. Varanasi[28] is smaller than Kāśī and constitutes the area between River Varana and rivulet Asi. Inside the territory of Varanasi, there is the Avimukta region and the innermost smallest area is called Antargṛha (inner sanctum).

Sukul (1974: 7) pointed out that Kāśī and Varanasi are not only political units but essentially sacred units. He introduced the term called *kṣetra* to differentiate between sacred and political units. Kṣetra, like the *tīrtha*, has several meanings and can be translated as a land, abode, sacred region, a place of pilgrimage, etc. The *kṣetra* is a spatial concept and frequently appears in Paurāṇika literature.

KĀŚĪ KṢETRA

Sage Vyāsa composed *Harivaṁśa* as an appendix to the *Mahābhārata. Harivaṁśa* narrates the story of Kāśa (he was the great grandfather's father of King Divodāsa) who ruled Kāśī, had many sons called Kāśīs. The kingdom ruled by Kāśa called as Kāśī (Mani 1975: 245).

Padma Purāṇa describes the region of Kāśī in a circular form placing Madhyameśvara[29] at the centre and the radius of around ten miles. The distance of this semi-diameter is measured from Madhyameśvara to Dehalī Vināyak.[30] Sukul (1974: 163) discussed the range from Madhyameśvara to Dehalī Vināyak (about 17 km.) situated on the western direction and expounded how the measurement of krośa differed in another Purāṇas and in the *Mahābhārata* that also discussed the realms of Kāśī *kṣetra.*

Figure 1.8: The text in this image informs about the starting point of the territory called Kāśī Viśvanātha Khaṇḍa
Notes: Photo by the author, February 2016

In chapter 57 of Kāśī Khaṇḍa, Śiva tells the locations of the fifty-six Vināyakas at Kāśī. These temples are arranged in seven concentric circles, and Dehalī Vinayak is located on the first outermost circle on the western side of Kāśī (Tagare 1996).

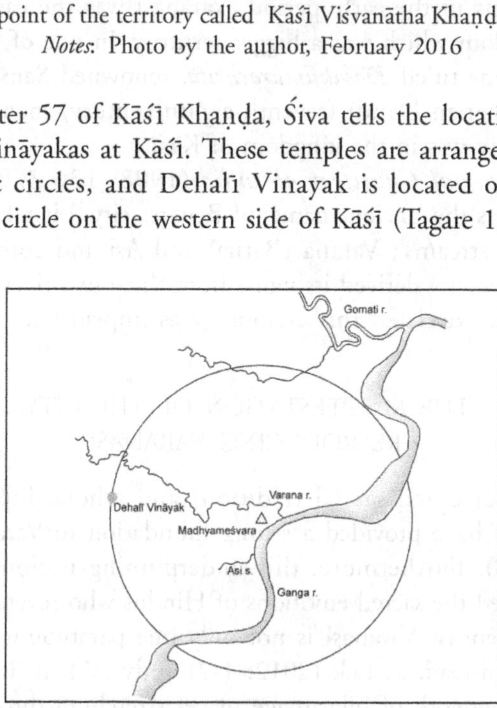

Figure 1.9: The circular formation of Kāśī *kṣetra* in a circular form showing the location of Madhyameśvara and Dehalī Vināyak on the western side.
Notes: The drawing is based on the map in Niels Gutschow, 2006: *Benares The Sacred Landscape of Varanasi*. Tracing and editing by the author

VARANASI KṢETRA

Kāśī Khaṇḍa mentioned Kāśikā as an earlier name and because of Varana and Asi, the city acquired the title 'Varanasi' (*SP*, KKh. 30.70). Kane (1953) and Sukul (1974, 1977) both cite Purāṇas to identify the territory of Varanasi. *Padma Purāṇa*, *Kūrma Purāṇa* and *Agni Purāṇa* and said unambiguously that Varanasi is located between the River Varana in north and the River Asi in the south. *Padma Purāṇa* also mentioned the River Ganga on the east and Pāśapāṇī Gaṇeśa Temple[31] (Sadar Bazar area) in the western direction.

The Purāṇas and other historical documents used both names, Varanasi and Kāśī in a synonymous way. However, Sukul (1974) drew attention to the fact that Varanasi and Kāśī were not used as synonyms in the early period. Varanasi was the capital of the Kāśī kingdom which was a bigger territory. In one of his poetical compositions titled *Daśakumāracarita*, renowned Sanskrit author and grammarian Daṇḍī (seventh-eighth century) mentioned that Varanasi city was in the kingdom of Kāśī.

The Imperial Gazetteer of India (1908: 136-7) stated that Varanasi was the 'ancient name' of Benares city.[32] It referred to the two 'small streams'; Varana (Barna) and Asi and continued that probably the city derived its name from these two rivers. However, the gazetteer declared this etymology as implausible.

THE MANIFESTATION OF THE CITY: REPRODUCING VARANASI

The uninterrupted sacred traditions and scholarship of Hindu philosophy have provided a strong foundation to Varanasi (Kane 1953: 618). Furthermore, this underpinning notion essentially encapsulated the sacred emotions of Hindus who revere the city as a living identity. Varanasi is not only one paramount sacred city of India. Instead, as Eck (2012: 1-2) aptly said, it 'is part of an extensive network of pilgrimage places stretching throughout the length and breadth of India'. The devout reverence for Varanasi reflects in the last will of Madhavrao Peshwa that was dated 30 September 1772.[33] Among the 9 points of this last will, three points mentioned Varanasi explicitly. The will asserted that all efforts should be made to bring back Varanasi and Prayag (formerly Allahabad), the two most sacred places that were under Muslim control.

Varanasi city along with the prominent Vishwanath Temple of Śiva also houses countless temples of deities of the Hindu pantheon, Sun worshipping places, several *kśetras* and *pīṭhas* of gods and goddesses. Simultaneously the city of Kāśī has been reproduced in all directions of the Indian subcontinent.[34] The city, its great Vishwanath Temple and sacred water pools are reimagined and reproduced throughout the Indian subcontinent. Reproducing Varanasi and its geography is an interesting phenomenon worth discussing. Almost all other reproduced Kāśīs are either situated on the celestial river bank or near the large lake or water pool and have reproduced the famous Vishwanath Temple. This phenomenon reveals the innate association between a pool and a temple.

CITY OF STREAMS

The landscape of Varanasi citys is encircled by river Ganga, River Varana and Asi stream. Purāṇas referred to these rivers and streams and described the peculiar forms, seasonal changes and commented on their cosmic characteristics. Kāśī Khaṇḍa states that Varana and Asi both rivers meet River Ganga which is flowing northwards. Kāśī Khaṇḍa also described the origin and chronology of the rivers. The text mentioned two streams called as Kiraṇā and Dhūtapāpā. These two streams then merge with Dharmanada. Later, Bhagirathi (Ganga) appears followed by Yamuna and Sarasvati rivers. The sacred confluence point of these five rivers is renowned as Pañca-Ganga or Pañcanada at Pañcagaṅgā Ghāṭa.[35]

GANGA APPEARED IN VARANASI

The Nadī Stuti Sūkta (river praising hymns) in *Ṛgveda* (10.75) enumerated the names of nineteen rivers commencing with River Ganga (Pandey 1949: 94). In *Mahābhārata*, when Dhṛtarāṣṭra asked Sañjaya to describe to him the geography of India (Bhārata),

Sañjaya presented a detailed picture of India before Dhṛtarāṣṭra who could not see. He enumerated the name of the mountains, rivers and the list of *janapadas.* While describing rivers, he said that inhabitants of this land drank the water of the following large rivers. The extensive list of rivers encompassing the entire geography of the Indian subcontinent begins with the name of the River Ganga.[36] The above selective two prominent references highlight the significance attached to one of India's celestial rivers Ganga. As Varanasi is the prototype of a *tīrtha,* the River Ganga is an epitome of sacredness (Eck 2015: 213-14).

River Ganga[37] is repeatedly mentioned as a northward-flowing river and revered as the most splendid *tīrtha.* Kāśī Khaṇḍa (5.18, 20) mentions that Varanasi is privileged with the sparkling water of River Ganga which is perpetual. However, the text reveals that the River Ganga was not always flowing in Varanasi and appeared or migrated later. One chapter of the text narrates the compressed form of the story of how Ganga appeared on earth.

King Bhagiratha bestowed a boon after performing hard penance and austerity and Ganga descended from heaven to wash the ashes of the king's forefathers. Shiva (principal deity of Varanasi) moderated the enormous speed and force of Ganga with his matted hair and thus Ganga appeared on the earth without causing destruction. The descendant of King Bhagiratha, Dilipa broguht Ganga to Varanasi where the Maṇikarṇikā (Cakrapuṣkariṇī) water tank already existed. After the confluence, with Ganga, the water of this tank also acclaimed greater sanctity.[38]

While expounding the etymology of Varanasi through the lens of river dynamics and the civilization that prospered in these vast flood plains, Raju and Pandey (2013: 138-9) claimed that 'Varanasi' possibly implied a city located on the banks of the Ganga; the 'ever-flowing river'. Due to significant tectonic events in this larger plain, this ever flowing River Ganga shifted eastward to Varanasi around forty thousand years back (Shukla et al. 2012: 110).

WATER IS OMNIPRESENT IN VARANASI

The Kāśī Khaṇḍa extensively listed the names of temples, the rituals to be performed at the temple, sacred bathing at the temple tank, ponds, lakes, or streams that are an integral architectural structure of the temple complex. The two entire sections of Kāśī Khaṇḍa offered the description and locations of more than 281 water reservoirs.[39] The list of all 281 titled water reservoirs is in the appendix section.

These types are categorized into

1. *tīrtha* (water pool),
2. *kuṇḍa* (square or rectangular-sized stepped water pool),
3. *puṣkariṇī* (pond with steps),
4. *vāpī* (well or an oblong reservoir of water),
5. *hrada* (deep-water place),
6. *dīrghikā* (long or oblong lake),
7. *sarōvara* (lake or large pond) and
8. *nālā* (rivulet).

The qualitative description of 281 water sites broadly covers the following attributes:

1. *Structural features*: deep pit, a vast lake, curative features of water.
2. *Geographical locations and settings*: direction and (qualitative) distance from the other water reservoirs in proximity or a temple site.
3. *Etymology*: former title, origin.
4. *Activities*: rituals, bathing, ablution.

The list of 281 water reservoirs includes the names representing the pantheon of Hindu gods and goddesses such as: Gaurī Kuṇḍa (77.47), Indra Kuṇḍa (97.116), Kali Tīrtha (84.75), Kedāra Kuṇḍa (77.54), Viṣṇu Tīrtha (61.144) and names of other pilgrimage places such as; Gayā Tīrtha (69.36), Prabhāsa Tīrtha (83.95), Prayāga Tīrtha (33.179), and Puṣkara Tīrtha (69.20).

The water reservoirs such as Cakrapuṣkariṇī Tīrtha (also called Maṇikarṇikā Tirtha), Kapālmocana Tīrtha and Matsyodarī Tīrtha are frequently cited in the context of their supremacy, virtuous features, merits and boons enclosed with the ablution ritual. All these three sacred water structures are still surviving despite continuous natural and human intervention. Tīrtha and kuṇḍa[40] are the most mentioned types in the entire text. As compared to *kuṇḍa*, the term *tīrtha* is a panoramic term, and with its philosophical context which has already been discussed, it can be referred to as any sacred water site.

Except for *tīrtha*, the remaining seven categories can be broadly sub-categorized into the natural and man-made types of water reservoirs. The *kuṇḍa, puṣkariṇī* and *vāpī* were the results of human intervention in the form of developing the supplementary structure for multiple purposes. Compared to these three forms of water reservoirs remaining *hrada, dīrghikā, sarovara* and *nālā* forms are relatively immune to human intervention.

Some of the titles of water tanks also highlighted the historical influence such as the Islamic invasion and demolition of temples closely connected with the water tank. The opening verses of chapter forty-seven of Kāśī Khaṇḍa described the worshipping of temple tank called 'Arka Kuṇḍa' near the temple of Uttarārk (Sun) located in the northern direction. The final verses of the chapter mentioned that this praiseworthy temple tank was also known as Barkari Kuṇḍa (popularly known as Bakariyā Kuṇḍa) (see Figures 1.10, 1.11 and 1.12). Tagare (1996; 1997) said that these verses were probably introduced in the later period and demonstrated the Islamic influence on the addition of a new title (Tagare 1996).

MATSYODARĪ: A RIVER OR A LAKE

Lakṣmīdhara in his treatise and the Kāśī Khaṇḍa referred to the Matsyodarī as a river (rivulet). The Kāśī Khaṇḍa described how the water of Matsyodarī was flowing into the deep ditch and a canal was built to bring the water of Ganga. The chapter gave the details of its internal and external divisions and merging with the water of Ganga and conferred the title of *tīrtha*. The subsequent verses from the chapter hinted at the etymology of this rivulet as it formed the shape of a fish (*matsya*) belly when the water of Ganga overflowed and entered the stream of Matsyodarī. (KKh. 69: 136-9).

In his book, *Kāśī kā Itihās* (*History of Kāśī*) Moti Chandra (1962: 2, 292) relied on the theory of Kāśī Khaṇḍa and described the two forms of Matsyodarī; one form was a lake (inner), and the other was that of a moat (outer). He also indicated that the parallel streams might symbolize the old Matsyodarī stream. However, Sukul (1977: 61, 248-9) discussed two additional seasonal streams; Pitāmahasrotikā (Brahmanāla) and Mandākinī with Matsyodarī. Sukul described this as a symbolic description of a 'natural phenomenon'. He explained that in the rainy season these three lakes used to transform into streams to regulate the excess water coming from the River Ganga and its tributary Varana. The metaphor of a fish belly shape of Matsyodarī in Kāśī Khaṇḍa delicately revealed how the water reservoirs used to adapt distinct forms depending upon the season to regulate the excess water of Ganga and her tributaries. Regarding the Matsyodarī stream, Sukul (1977: 246) made a claim that probably in the fifteenth century the stream had vanished or disappeared.

The mapmaker in the pictorial map of Varanasi (labelled as 'Map of Benares'[41]) probably depicted this natural phenomenon by depicting the stream of Matsyodarī flowing parallel to Ganga and its confluence with rivulet Varana in the north direction (see Figures 1.10 and 1.11). However, the mapmaker also showed Matsyodarī in a rectangular stepped tank and the water of tank

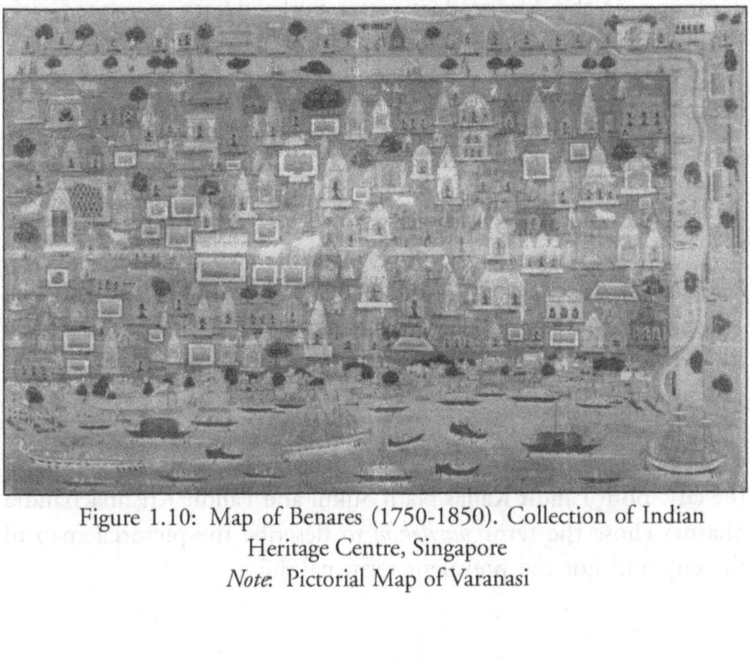

Figure 1.10: Map of Benares (1750-1850). Collection of Indian Heritage Centre, Singapore
Note: Pictorial Map of Varanasi

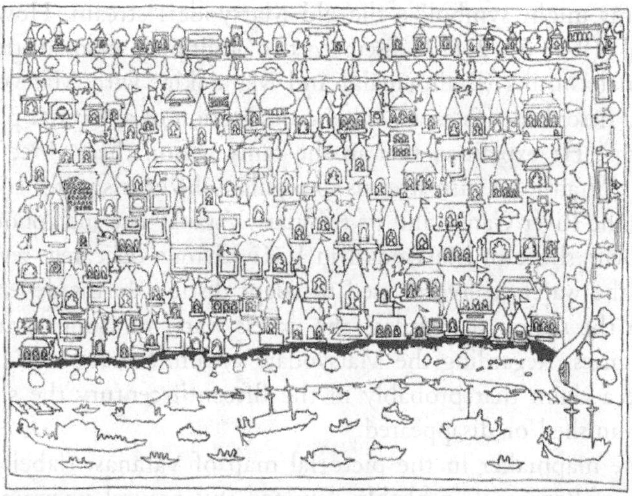

Figure 1.11: Tracing of pictorial map of Varanasi (Map of Benares)
Note: Highlighting the possible stream course of the Matsyodarī stream (Drawing by the author)

in an oval form with a corrupted title 'Machhodarī *tīrtha*'. The depiction of the Matsyodarī water tank and its corrupted title Machhodarī help one to verify the period of this pictorial map.

LOCATING SACRED SITES AND SPATIAL TEXTS

Swami Medhasananda (2002: 170-1) recounted the studious exercise of two Baṅgālī scholars during the eighteenth century to locate the forgotten sacred sites with the help of spatial texts. Medhasananda also narrated the stupendous efforts of Ramakrishna Dikshit Gor,[42] who referred extensively to the text of Kāśī Khaṇḍa to identify the ancient temple sites which are in decay or lost and endeavoured to restore them. Corresponding to this exercise, two other scholars of Varanasi carefully studied the text of Kāśī Khaṇḍa to produce a *parilekha* (an outline or a sketch without colour) of the city. Both Pandit Kailas Nath Sukul and Pandit Krishnachandra Sharma chose the term *parilekha* to describe the pictorial map of the city and not the prevalent term nakśhā.

Pandit Kailas Nath Sukul published one of the most detailed maps (see Figure 1.12) of the city in September 1876 (Gutschow 2006: 56). Sukul offered a brief description of his pictorial map titled 'Kāśī Darpaṇa' (mirror of Kāśī).[43] He expounded that this map depicted titles of deities and sacred sites mentioned in various Purāṇas, including *Skanda Purāṇa*. Gutschow (2006: 55) enumerated more than 1,200 entries on this map. The map represents the substantial drawing of some of the prominent temples and modern edifices such as 'town hall'. It also maps the river, streams, temple tanks and wells carefully. In the initial assessment, Gutschow argued that this map does not operate within the boundaries of a topographical representation. This map represents some of the places which had lost their terrestrial nature. However, when Gutschow juxtaposed Sukul's map of 1876 with the topographical map surveyed in 1928-9, he observed the correspondence in the spatial positioning of temple sites, water pools and modern edifices (Gutschow 2006: 70-1).

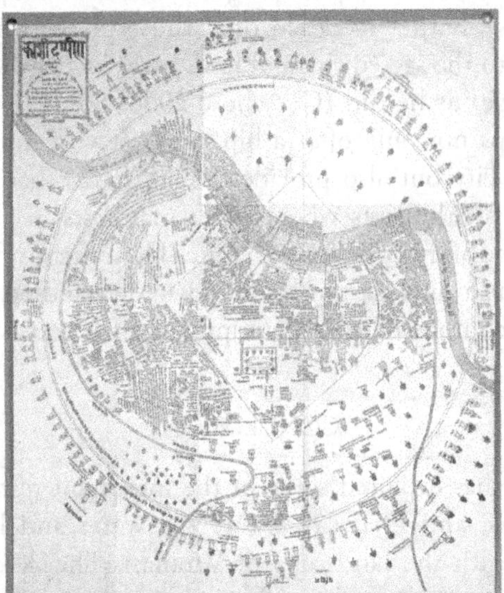

Figure 1.12: Kāśīdarpaṇa by Pandit Kailas Nath Sukul, 1876. Printed at Vidyodaya Printing Press, Varanasi. The British Library, London
Note: The map is oriented towards the east

Sukul's map of 1876 tried to restore some of the water sites and streams that were extinct many years ago, even before the map was published. Sukul relied upon and followed the order of Kāśī Khaṇḍa to record all the *tīrthas* and water pools and (re)arranged the natural settings (Michaels 2006). Axel Michaels (2006: 134) studied the map and called the Sukul map of 1876 a concise thematic form of Kāśī Khaṇḍa.

One year later, in 1877 Pandit Krishnachandra Sharma published the map (contains three parts) titled 'Kāśīdarpaṇapūrti' at the same printing press. While describing the details of the map in the colophon, Sharma writes that Kāśī Khaṇḍa and another Purāṇas, Nibandhas are referred. Additionally, He visited all the locations and presiding deities under the guidance of Ramakrishna Dikshit Gor. He also mentioned that this *parilekha* was produced following the rules of European Cartography prevailling at that time.

Gutschow (2006: 74) pointed out that all three maps were 'drawn to scale'. While analysing the 1877 maps, Gutschow compared it with the one-year-old map and argued that 1877 maps were more 'scientific' as Sharma recorded only those places where pilgrims could travel to the sacred site and have a *darśana* of the *vigrahaḥ* or *mūrtiḥ* (image) as he did (Gutschow 2006: 80).

These maps not only offer a bird's-eye view of the sacred landscape of the city but also provide the apparatus that facilitate the pilgrim to visit the sites where water manifests in various forms reminding one of the cohesion of nature. The exercise of constructing these maps also disclose the profound understanding of the composers who composed these descriptive spatial texts.

CONCLUSION

This chapter has explored some of the important terms such as—sacred, *tīrtha*, and *kṣetra* to locate the water and its profound connection with the landscape of Varanasi. The descriptive texts recognise the centrality of free flowing water in various forms and how it preserves the sanctity of the place. This chapter is devised to reconstruct the pre-colonial landscape of the city and its comprehensive approach reflected in the descriptive texts to perceive

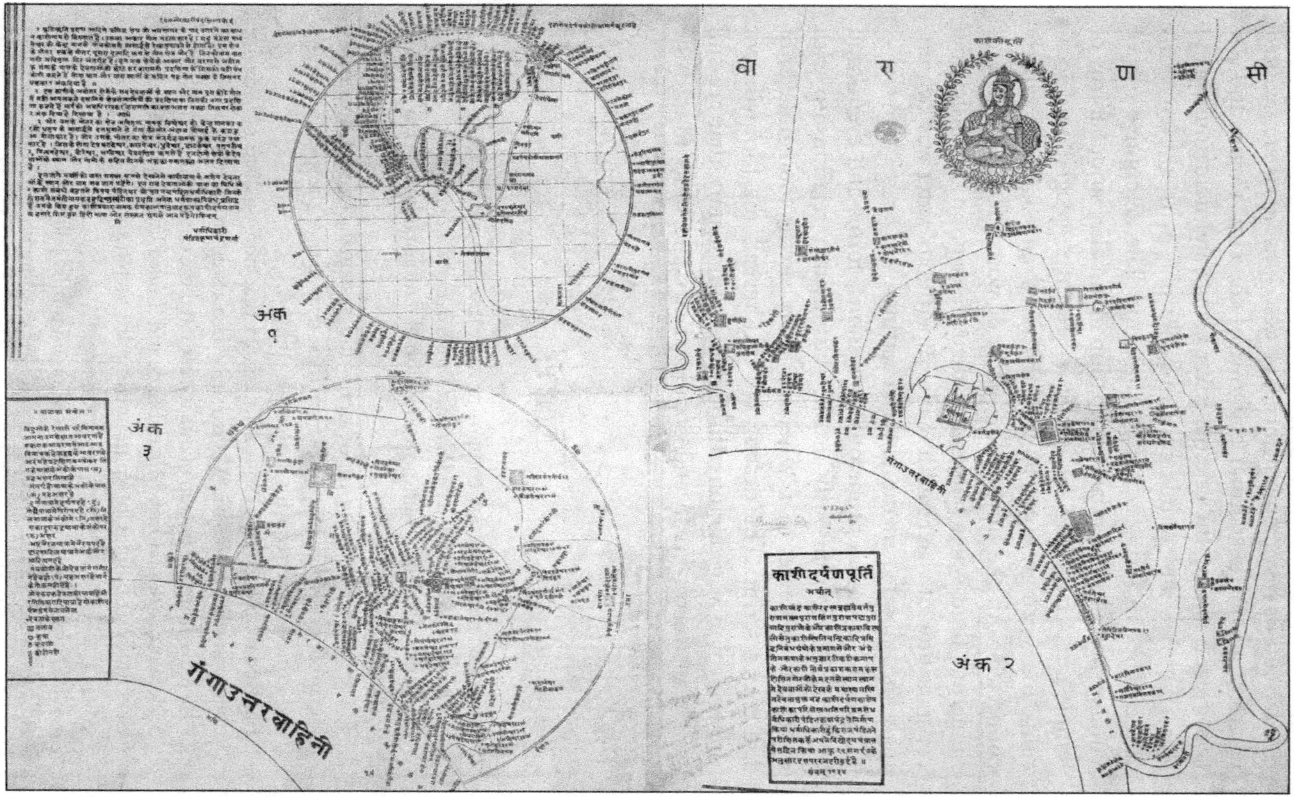

Figure 1.13: Kāśīdarpaṇapūrti by Pandit Krishnachandra Sharma, 1877. Printed at Vidyodaya Printing Press, Varanasi.
Courtesy: British Library.

the dry and its fluid components. The descriptive texts are revisited to comprehend the natural environs of the city. Varana and Asi two streams joining the celestial River Ganga and numerous inland water courses and ponds continuously play with the environs and coalesce to form the sacred landscape. The descriptive text of Purāṇas provides essential information about the various forms of temple tanks regarding the 'structural features, geographical locations and settings, etymology of the titles' and activities closely connected with the surrounding pools of water. The repository of numerous fluid water bodies and their description establishes the constant flow of water and how the superfluous water was swimmingly incorporated in the physical settings.

Kāśī Khaṇḍa has numerous expressions, forms and events and actions that continuously invoke the intrinsic relationship with water and reiterates that 'water sensitivity' as the fountainhead of the entire script. The descriptive text reveals the footprints of the pre-colonial geography through its commentary that are still visible in today's urban and populated city and hints toward the 'of' rubric.

NOTES

1. Nibandha means literary composition, treatise or a compendium. Vaman Shivaram, Apte, 1890, *The Practical Sanskrit-English Dictionary.*
2. It also describes the profound association of temple tanks with daily rituals and pilgrimage practices.
3. These texts record the topography and seasonal changes, mostly in metaphorical forms. The verses from these texts narrate the specific features and stories connected with the temple tanks.
4. Baidyanath Saraswati (1984: xvii) argued that non-dualism or 'Advaita' philosophy differs from the Western social scholarship on the study of religion, which discusses the notion of duality. He cited the contribution of Nathan Söderblom, Rudolf Otto, Mircea Eliade and Émile Durkheim to discuss the classification of things into pure and impure, sacred and profane, good and evil.
5. Kāśī Khaṇḍa (84.53) stated that Varanasi is full of all types of *tīrthas.*
6. Ādi Śaṅkara describes Varanasi city (Kāśī) as a supreme *tīrtha* and identified

it with the physical body. In one of his compositions titled *Kāśīpañcakam*, which is a compact style of five verses hymn (*stotra*) Ādi Śaṅkara mentioned the other sacred places such as Maṇikarṇikā and River Ganga.

7. Vaman Shivaram Apte, 1957-9, *The Practical Sanskrit-English Dictionary*. Kuber Nath Sukul (1977: 59) classified *tīrthas* into *sthāna tīrtha* and *jala tīrtha*. *Sthāna tīrthas* are some famous sacred sites whereas rivers, lakes, water pools wells, come under the category of *jala tīrtha*.
8. Compared with the Bhandarkar Oriental Research Institute edition of 1942, the Kumbakonam edition (based mainly on the south Indian texts) of the *Mahabharata* titled *Srimanmahabharatam* (1906: 82, 77-8) however mentions the Avimukta Tīrtha.
9. Sage Agastya to believed to have travelled to different parts of India and build hermitages. His name appears in the *Vālmīki Rāmāyaṇa* and Tamil literature. He composed *Agastya Saṃhitā* in the *Skanda Purāṇa*. For more information on Sage Agastya, see Mani (1975).
10. *Vāyu Purāṇa* significantly discusses water, its nature and the water cycle. The text explains that funeral rites or ceremonies (*śrāddha*) should be performed near—sacred lakes, rivers, pilgrimage places, mountains, etc., for receiving all the proper virtues at the right time. (*VP* 15.2-3; 42.1; 51.16).
11. Julia Hegewald (2002: 91) referred to the term tank temples and described them as specially reserved for religious ceremonies and not for everyday usage. However, in Varanasi city, most of the tanks are now used for various purposes including temple rituals, immersion of images, ablutions, bathing and washing.
12. Fieldwork and interviews from August 2016 to November 2019.
13. Members regularly commuting to *akhāḍās* pointed out the waning condition of *akhāḍās* due to changes in lifestyle and emergence of the modern culture of sports gym. They talked about how *akhāḍās* are losing their popularity. During the fieldwork, the members also referred to the failing link which used to bond the temple tanks and *akhāḍās*. The deteriorating condition of *akhāḍās* is also responsible for the poor maintenance of temple tanks and ponds.
14. The eighteen Purāṇas, are 1. *Brahma*, 2. *Padma*, 3. *Viṣṇu*, 4. *Vāyu*, 5. *Bhāgavata*, 6. *Nāradīya*, 7. *Mārkaṇḍeya*, 8. *Agni*, 9. *Bhaviṣya*, 10. *Brahmavaivartta*, 11. *Varāha*, 12. *Liṅga*, 13. *Skanda*, 14. *Vāmana*, 15. *Kūrma*, 16. *Matsya*, 17. *Garuḍa*, 18. *Brahmāṇḍa*. However, scholars have differences on the total number of Purāṇas and their order.
15. Ludo Rocher (1986: 131) highlights the importance of Purāṇas in the context of understanding the pre-modern geography of India.

16. *Skanda Purāṇa* is a repository of all the Sthala Purāṇas and Tīrtha Māhātmya (Tagare 1992).
17. S.M. Ali (1966: 13) discusses the geography described in Purāṇas and mentions the *Bhuvanakośa* as a treasure of terrestrial mansions. He also refers to Bhuvana Sāgara (ocean of terrestrial mansions or inhabited lands), Bhuvana Khaṇḍa (section of the earth), Trilokya Darpaṇa (mirror of the worlds) and Kṣetra Samāsa (combination of countries). Joseph E. Schwartzberg (1992: 312) describes the Bhuvanakośa as a 'geographical list' and equates it to a proto-gazetteer.
18. The other six chapters (or sections) of *Skanda Purāṇa* are: Maheśvara, Viṣṇu, Brahma, Āvantya, Nagara and Prabhāsa.
19. The description of merits and glories of a sacred place.
20. The seven chapters of *Skanda Purāṇa* are (1) Maheśvara, (2) Viṣṇu (Vaiṣṇava), (3) Brahma, (4) Kāśī , 5. Āvantya, (6) Nāgara, (7) Prabhāsa.
21. Hans Bakker, 2009, *Rama Devotion in a Saiva Holy Place: The Case of Varanasi.* (p. 72). Rajendra Chandra Hazra (1940: 7) draws one's attention to the fact that the structure of Purāṇas was dynamic and the addition of new chapters from other Purāṇas and Smṛti was a common exercise to enrich the latter work.
22. Maya Prasad Tripathi (1969: 178, 321) discussed the term 'Bhuvanakōśa' as corresponding to modern geography.
23. Kuber Nath Sukul (1977) explained that the list of 340 temples included only the most prominent temples and other sacred places.
24. Kamalakrishna Smrititirtha (1912) mentioned Vācaspati Miśra who composed *Tīrthacintāmaṇi* was from the city of Mithila and during Raja Harinarain Deb's rule.
25. Dilip Da Cunha (2019: 234) develops an interesting perception of *janapada* around the dynamics of 'high grounds that anchored in an ocean of rain'.
26. A commentary on Pāṇini's *Aṣṭādhyāyī* (grammar) by Vamana and Jayaditya.
27. Sage Agastya asks Skanda to explain to him why the city was called by different names: Kāśī, Varanasi, Rudravāsa (abode of Rudra; one of the manifestations of Śiva or Śaṅkara), Avimukta, Ānandakānana (forest of pleasure), Mahāśmaśāna (great cremation ground), in different times.
28. The present name of the city 'Varanasi' was officially declared in 1956 by the Government of India. Varanasi city is the administrative headquarter of Varanasi district.
29. Kāśī Khaṇḍa described the Madhyameśvara as the focal point. (*SP* KKh. 33.170)
30. Mitramiśra (1610-40) composed *Tīrthaprakāśa* (part of Vīramitrodaya)

and excerpted the verses from *Padma Purāṇa.* The verses from *Tīrthaprakāśa* mentioned *Madhyameśvara* and *Dehalī Vināyak* to define the territory of Kāśī.

31. Pandurang Vaman Kane (1953) mentioned about that Dehalī Vināyaka temple was in the western direction.
32. *The Imperial Gazetteer of India,* Provincial Series: United Provinces of Agra and Oudh, vol. II, 1908.
33. The third point of the last will elucidated that reclaiming the sacred places such as Varanasi and Prayag from the Muslim domination was desired and endeavoured by the forefathers of the Peshwas.
34. The numerous sacred places, temples and monasteries situated in all directions of India along with its rivers are clustered and articulated in the numerical representation form of three sacred sites pilgrimage (*tristhalī*), four sacred abodes (*cāra dhāma*), five Kāśīs (*pañca kāśī*), seven; sacred cities (*saptapurī*), Gangas (*sapta-ganga*); celestial rivers (*sapta puṇya nadī*), eight Vinayakas (*aṣṭa-vinayaka*), ten sun places (*bhāskara kṣetra*), twelve *jyotirliṅgas* (*bārā jyotirliṅga*), etc.
35. Ganesh Vasudeo Tagare, 1996. (*SP* KKh. 59.109-15). Kuber Nath Sukul (1977: 60) claimed that both Kiraṇā and Dhūtapāpā are underflowing streams flowing parallel to River Ganga on the western side.
36. The Critical Edition of *Mahabharata.* Bhīṣma Parva 10.13. Suryavanshi (1986: 120-1) provides lists from two chapters of the *Mahabharata.*
37. Chapter twenty-nine of Kāśī Khaṇḍa a titled as 'A Thousand Names of Ganga' (*Gaṅgā Sahasranāma*) lists the names of Ganga in the form of the Goddess. Tagare (1996, 1997) mentions that the list of thousand names contained 997 names and gives the explanation that the list also included the epithets in the verses. The list included titles such as: River Alaknanda (a tributary of Ganga), extremely sacred (*pūtā*), *tīrtha,* river with rapid current (*vegavatī*), heavenly Ganga (Mandākinī), flowing slowly (*laghuvegā*), etc.
38. Based on the English translation by Tagare (1996, 1997), the text was paraphrased adding additional information for more clarity.
39. This chapter excludes the category of *vāpī* in the list of water reservoirs of Varanasi.
40. Julia A.B. Hegewald (2002) in her research on the water architecture of South Asia discussed tank and *kuṇḍa* as two main categories of water reservoirs found in India. She distinguished these two types based on their building techniques, surface area, the formation of steps and size. She defined a tank with relatively large surface area, short heightened steps with little depth and rectangular or square form. According to her, *kuṇḍa* were

small- sized funnel formed and had steep steps. She highlighted the vital observation that, however, the local types attached with these water reservoirs were not necessarily identical in their structural design.

41. This map is probably a part of a larger pictorial map of Varanasi. The *ghāṭa* (masonry steps on banks of a river or water tanks) depicted on the furthermost left side of the map is Bindu Mādhava Ghāṭa (also known as Veṇī Mādhava Ghāṭa and part of Pañca-Gaṅgā Ghāṭa). After that, Durgā Ghāṭa, Brahmā Ghāṭa, Gayā or Gai Ghāṭa, Trilocana Ghāṭa, Pralhāda Ghāṭa and Rāj Ghāṭa are mentioned.
42. Reverend Matthew Atmore Sherring (1868: 105-6) mentioned the meticulous work of locating the obscure temple sites and consecrating the figures of deities there by Ramakrishna Dikshit Gor (ji). Contrary to his notions, surprisingly Sherring appreciated the work and called it 'reviving Hinduism' carried by a 'Gujarati Brahman'. Ramakrishna Dikshit Gor also published the list of temples titled 'Kashiyatraprakasha'.
43. This map was studied profoundly and investigated by the 'Varanasi Research Project of the South Asia Institute' at Heidelberg University in 2001.

CHAPTER 2

The Tale of Two Maps

This chapter is a continuation of the previous chapter and corroborates the descriptive texts with thematic mapping. It first examines how the celestial qualities described in literary works have shaped and produced the earthly forms of sacred places. It tries to analyse how these spatial texts subtly influenced the process of producing pictorial maps of Varanasi. In doing so, the indigenous pictorial maps are juxtaposed with the Western, scientific, topographical maps produced primarily by colonial officers during the colonial time.

While analysing the two distinct spatial representations of Varanasi, this chapter develops the critical argument that notions of Western cartography and practices have rejected the Indian way of understanding space and its distinct representational approaches primarily. The map-making tradition in India arguably visualizes space symbolically and conceptually that does not comply with the yardsticks of modern map-making practices. This chapter, therefore, re-examines the pictorial maps in the light of archaeological and geological investigations to reveal how the mapmaker possibly hints at the peculiar topographical formation of the city.

In 1784, Jesuit father Joseph Tieffenthaler surveyed and published the map of the courses of Ganga and the Ghaghara (Karnali) rivers. Since then, the River Ganga and the Varanasi region were continuously surveyed and exemplified in the forms of topographical maps made by colonial surveyors. Before the eighteenth century, topographic maps of Varanasi city were not accessible. However, as discussed in the previous chapter, the thorough reading of Paurāṇika texts and treatise composed and recomposed in different periods reveal the concealed layers of the geography of the city.

The first topographic map was produced by British scholar James Prinsep (1799-1840). Within the first two years of his tenure in the office, he surveyed the city and produced an accurate map out of his amusement. This map was titled 'The City of Bunarus, surveyed by James Prinsep' and published in 1822. The James Prinsep map of 1822 in several ways typified how the early colonial authorities came to organize notions of space and geographical order. The British cartographic notion of space, however, as will be pointed out, was markedly different from that of an indigenous conceptualization, as evidenced in the recent discovery of the 'Map of Varanasi, Shiva's Sacred City' that was drawn probably in the late eighteenth century by an unknown mapmaker. This map of Varanasi is relatively a recent discovery and was earlier catalogued by a somewhat misleading title.[1] This 'schematic map' earlier had a label describing it as a 'pictorial chart' and not a map.[2] Probably this work is acknowledged and evaluated more as a painting and not explicitly as a 'map'.

Harley and Woodward (1992: xvi) defined maps as graphical representations that primarily provided the spatial understandings of comprehensive dimensions of the 'human world' after analysing the etymologies of Latin and other European words for the map. This comprehensive approach marked a significant departure from the notion that restricted the approach of map-making processes and rejected the indigenous maps.

These two sharply contrasting maps of Varanasi can reveal, the marked differences over how the city was understood as an ecological formation. The careful analysis of two maps uncovers how the map makers of these two maps visualized the city located on a prominent river and had numerous water pools.

COLONIAL MAPPING EXERCISE

The surveying exercise of Varanasi goes back to 1765 when a Jesuit missionary named Joseph Tiefenthaler (1710-85) visited Varanasi and conducted same survey of the course of the River Ganga. Later, between 1784 and 1786, Captain Brown of East India Company carried out a survey. After that, from 1788 Capt. Francis Wilford who was a serving officer in Bengal Engineers continued the survey of the Varanasi province. Wilford was closely associated with Sanskrit College at Varanasi. Wilford studied Purāṇas and other literary texts with the assistance from a Hindu *paṇḍita* (scholar) to understand the Hindu cosmography and geography. In this lengthy essay, he asserted that Hindus lacked the notion of history and geography (Wilford 1809: 267, 269). He further added that even though Hindus had world maps and maps of Indian towns, these maps were not drawn by applying a proper scale and coordinates. The rejection of the Hindu notion of geography and carto-literacy emphasized by Wilford arguably overshadowed the future research on geographic understanding and map-making practices of India.

In the book titled *History of Cartography*, scholar Leo Bagrow (1964: 207) argued that India is a 'closed country' had (a) little knowledge of geography and (b) dearth of cartography. In his book, he defined a map as a scientific document and map making as an art of measurement, which required drawing with a high level of precision. 'Scientific document' and 'precision' are two important notions that have mainly shaped the western map-making practices and become the benchmarks to evaluate any map or a chart. Possibly map making went through multiple phases and was frequently modified by the new information and invention of new tools and devices. The oldest surviving map from the European continent was found in the Netherlands, dated the beginning of the fourteenth century and was a simple representation of places with notes (Harvey 1987: 470).

The travelogue of Thomas Twinning published in 1893 is probably an apt example of how the Western mind was perplexed to observe things like astronomy, almanacks, algebra, etc. Twining (1893: 256) wrote that he had purchased an accurate map of Delhi (city) along with some other places and an almanack. In one instance, Twinning mentioned the detailed direction he had received from local people about the routes. However, during his expedition in northern India, Twinning (1893: 301) referred to the map of Rennell and found that the position of 'Ally-Gunge' (Aliganj) marked by Rennell was incorrect. Further, he stated that though, Hindus knew about astronomy, algebra and mathematics they lacked the knowledge of geography and medicine.

Joseph Schwartzberg (1992, 296) said that previous studies on geography and cartography had either overlooked or ultimately rejected the available corpus of South Asian cartography. He referred to the book *Indian Maps and Plans* by Susan Gole and acknowledged her work of detailed research and collection of indigenous Indian maps, charts and plans. Jörg Gengnagel (2011: 13) alleged that Western interpretation of the history of Indian subcontinent cartography crafted a hypothesis that maps of Indian subcontinent were made only by Western cartographers.

Besides the humid tropical environment, which caused severe trouble in storing the maps, plans and charts, many maps were often discarded by colonial rulers after their usage (Schwartzberg 1992: 327). This common practice perhaps resulted in the scarcity of Indian maps that compelled the historians and researchers to either dismiss or make a cursory reference to the history of map making in India. Additionally, the accessible indigenous maps were primarily evaluated by a modern standard such as scale, proportional representation and orientation.

Western cartography practices were a combination of many objectives. Thomas Bassett (1994: 316, 326) discusses the European mapping and colonizing activities in West Africa. He stated that depicting the region as a part of the European colonies was effectively to collect the acceptability and patronage for the colonial ventures. Surveying and mapping the foreign territory was to augment the supremacy of European dominance. Kapil Raj (2007: 77-8) and Michael Wintle (2016: 278) have both discussed the unique cartouche in Rennell's map of 1782 (see Figure 2.1) to explain the support Europeans received from the local scholars.

In this symbolic depiction, the Britannia is standing high on the steps of a masonry statue of a lion, who is clutching a globe. The three learned people are standing with all humility, are offering the documents written as 'Shaster' (probably the Sanskrit word 'Shastra') to Britannia. On the top of the outer decorative circle, the sword and a rod of Asclepius are arranged to cross each other in the background of opium poppies (Raj 2007: 77).

Manu Goswami (2004: 157) argued that before the induction of Western maps, there was no equivalent word[3] for the 'map', especially in the Sanskrit language.[4] She further made the critical

Figure 2.1: Detailed view of the cartouche on James Rennell's *Map of Hindoostan*, 1782

comment that the maps from Paurāṇika works lacked proper scaling and were strangely oriented towards the east. She stated that modern Western cartography does not allow to manifest the territory in a symbolic and ambiguous form; instead, it combines diverse regions into one single uniform space. However, Harley and Woodward (1992: xvi) elucidated that the origin of the word map in Latin was 'Mappa' or 'Carta', meaning cloth or a formal document respectively. Dineschandra Sircar (1960: 246) mentioned the common word *nakśā* (or *naqśā*) which comes from the Arabic word has several connotations other than a map. He highlighted the word *mānacitra* denoting map, chart, plan, diagram, etc. Raj

(2007: 85) drew attention to the fact that many maps of Indian subcontinent made by European mapmakers before Rennell's map (which was first published in 1782) were drawn with varying scales and orientations.

The orientation of a map was influenced mainly by the prevailing cosmographical notions, the favourable direction of rulers and the faiths of that territory (Birch 1949: 3). John Short (2003: 14-15) explained the Latin word 'orientation', meaning east. He referred to the medieval period when mapmakers customarily oriented towards the east where the Holy Land (Jerusalem) is located. The Church preferred eastward orientation, however, others such as Arab and Roman maps were drawn facing the south. Most of the Maratha maps adhering to the Hindu tradition were also drawn placing east at the top (Gogate et al. 1998: 139). The English maps started orienting maps toward the northern direction which happened in the middle of the fifteenth century and that probably influenced the land compass (Harvey 1987: 473).

THE FIRST 'ACCURATE' TOPOGRAPHIC MAP OF VARANASI

In 1820, James Prinsep (1799-1840) arrived in Varanasi as Assay Mint Master and started surveying Varanasi out of his interest for two years. He then published his map of Varanasi titled 'The City of Bunarus, surveyed by James Prinsep 1822' (see Figure 2.2).[5] The map of 1822, satisfied all the prevailing Western cartographic norms and was referred to as the first accurate topographical map of the city. The scale of the map was 8 inches to the mile (approx. 1.609 km). At the bottom of the map, there is a north pointer and a slightly tilted vertical line to show the accuracy of the north direction. Prinsep illustrated an exceptional sketch of one of the Hindu temples placed on the *ghāṭa* of River Ganga as a cartouche.

The 1822 map provides information about Varanasi city, such as roads, alleys, brick houses, mud houses, *muhallā* (ward), *bāzāra* (marketplace), etc. While working on this topographical map, Prinsep also carried out a census of the city. His census provides a

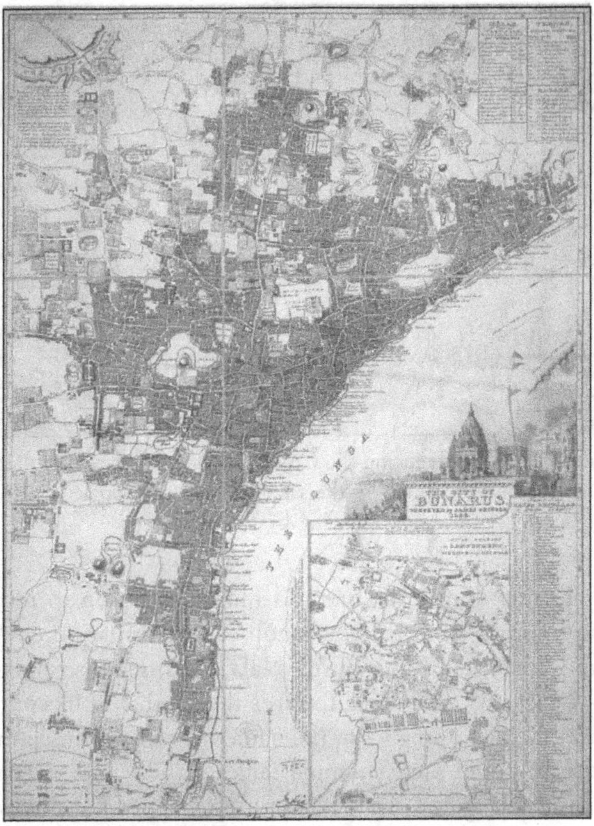

Figure 2.2: The City of Bunarus, surveyed by James Prinsep, 1822, The British Library Board, London

list of 1,000 Hindu temples, 333 mosques (Sherring 1868) and 174 gardens (Singh 1993).

The 1822 map reveals some of the meticulous survey work carried out by Prinsep. The map of 1822, with fewer variations, can be georeferenced on the satellite imagery (see Figure 2.3) and thus it depicts the detailed observations and surveys made by efforts carried by Prinsep before two centuries.

Prinsep mentioned the names of all *ghāṭa* (masonry stairs on the riverbank) beginning from the south towards the north on the west side of River Ganga. The map also illustrates the River Ganga

Figure 2.3: Georeferencing the City of Bunarus, surveyed by James Prinsep of 1822 map on the satellite imagery–Google Earth
Note: Drawing by the author

and its channel during the dry and monsoon season to reveal the variation in water level. In the left top corner, Prinsep drew Varana River and labelled it 'The Burna Nālā'.[6] Prinsep also mentioned the Asi rivulet as 'Ussee Nālā' towards the south and its confluence with River Ganga. Along with, the names of *ghāṭa*, Prinsep mapped the entire 61 titled sacred temple tanks, ponds[7] and several wells of Varanasi (see Figure 2.4). When Prinsep mapped the city, possibly the Agastya Kuṇḍa and Bula Nālā were extinct. However, Prinsep marked the possible location of these two water reservoirs.

LAND AND WATER SEPARATION

The map of 1822 surveyed by Prinsep is a land-centric representation that followed all the cartographic yardsticks and thus enjoyed the reputation of the first accurate and modern map of the city. It reveals all the information that its reader who is conversant with basic principles of cartography seeks. The map provides information such as street patterns, types of housing structures, gates, etc., based on the actual survey carried by Prinsep.

Pieper Jan (1979: 215) compared Prinsep's map of 1822 with

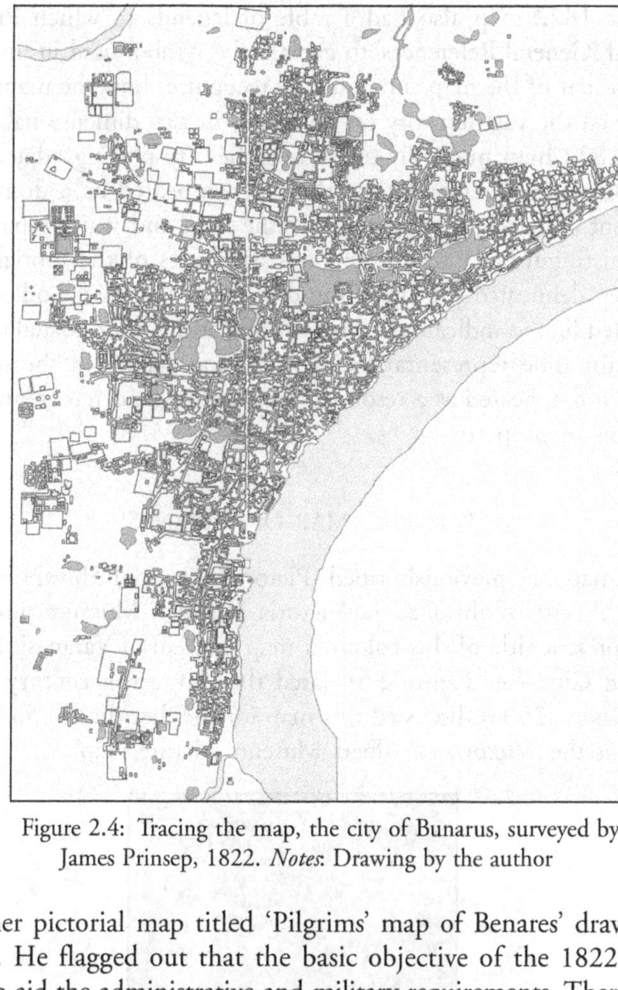

Figure 2.4: Tracing the map, the city of Bunarus, surveyed by James Prinsep, 1822. *Notes*: Drawing by the author

another pictorial map titled 'Pilgrims' map of Benares' drawn in 1875. He flagged out that the basic objective of the 1822 map was to aid the administrative and military requirements. Therefore, the 'public accessibility' and 'general significance' of the place was displayed effectively without concerning the sanctity of the place.

Prinsep's map of 1822 empowered the colonial administration to impose the property tax, construct roads and railway lines and introduce the central water supply. The detailed map allowed the regime to control the movements of people to handle emergencies. Since the map minutely plotted the spatial details, it is referred to as a base for subsequent mapping exercises.

The 1822 map also had a table of legends to which Prinsep tagged 'General References' to explain the symbol used in his map. The reader of the map consequently recognizes how the mapmaker confined the vast territory on the paper in two-dimensional form. The 1822 map markedly embodied the city placing adjacent to the principal river and thus the river appeared as a dominant element. The map revealed subtly the land and water separation and distinguished the dry and wet attributes of the topography. Prinsep delineated the area around some of the water bodies with a dotted line to indicate that this specific area got inundated during the rain. The representation of water was purely in the secular form and appeared as a resource, whereas the land represented the notions of property.

PICTORIAL MAP OF VARANASI

This map was previously titled 'Pictorial Chart of Shivapura. The heavenly city of Shiva' at the Victoria & Albert Museum, London. The present title of this coloured map is 'Map of Varanasi, Shiva's Sacred City' (see Figure 2.5) dated the eighteenth century. Niels Gutschow (2006) discussed this map for the first time and referred to it as the 'Victoria & Albert Museum picture map'.

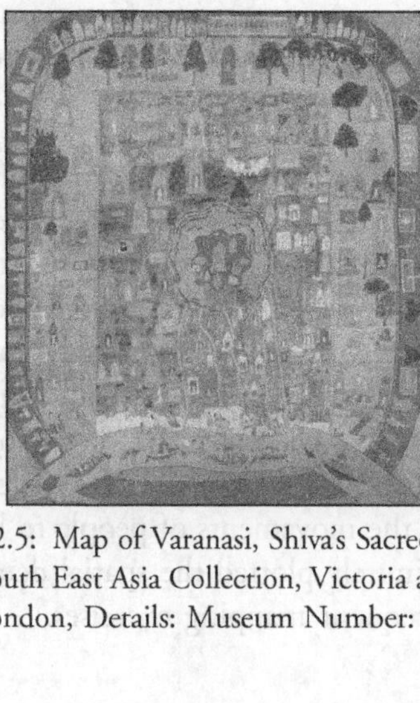

Figure 2.5: Map of Varanasi, Shiva's Sacred City
South & South East Asia Collection, Victoria and Albert Museum London, Details: Museum Number: 09322(IS)

However, Gengnagel (2011) proposed the title 'Stylized map of Varanasi'. The author proposes the name, 'Pictorial Map of Varanasi' and after interpreting the basic layout of the map will elaborate on the water structures of the map.

The pictorial map of Varanasi was westward oriented, and the River Ganga was depicted on the bottom side, i.e. to the east side. River Varana was depicted on the right side (north) and Asi rivulet on the left side (south) of the map.[8] The temple of Viśveśvara or Viśvanātha and the main complex was located in the centre of the map surrounded by yellow coloured eight circles with smooth curves on eight directions which constituted the Antargṛha Kṣetra (innermost sanctum circuit). Immediately outside the eight central circles, enclosed by thin line was the area referred to as Avimukta Kṣetra (the place never forsook by Śiva; the principal deity of Varanasi). Also there were five streams flowing down from the Antargṛha Kṣetra and joining the River Ganga at the bottom.[9] The area outside of the Avimukta Kṣetra is referred to as Kāśī and the outermost area is called Pañcakrośī Mārga (pilgrimage route) (Gengnagel 2011: 74). The map provides neither any navigational information nor the detailed location of the temples and water pools. No paramount importance or preference has been given to any temple. The depiction of Kāśī Viśvanātha Temple is also not exceptional except it is depicted in the centre of the map (see Figures 2.6 and 2.7).

Gengnagel (2011) give a detailed description of the temples and temple tanks and described the map in five sections based on the boundaries of the territory. To discuss the spatial distribution and arrangement of forms, one focuses on mainly the sacred temple tanks and ponds depicted on this map. These temple tanks and ponds are classified into nine categories based on their features and structural forms (see Figure 2.7).

CATEGORIES

Group 1: River and rivulets: (1) River Ganga, (2) Varana Saṅgama (confluence of River Varana with River Ganga), (3) Asi Saṅgama (confluence of Asi rivulet with River Ganga).

Figure 2.6: Tracing of all the forms on the Map of Varanasi, Shiva's Sacred City
Note: Drawing by the author

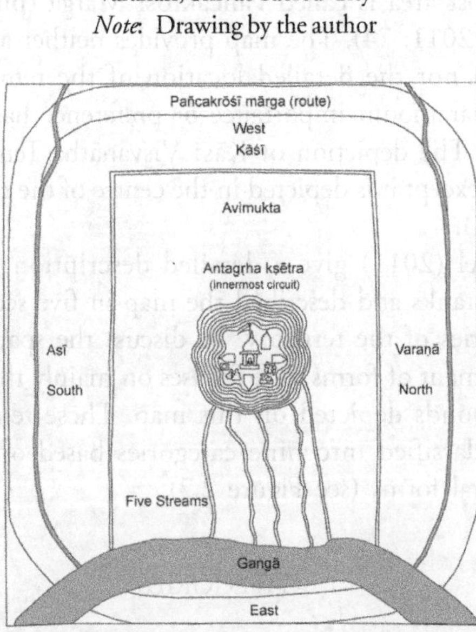

Figure 2.7: Tracing of main zones and their titles (including river and rivulets) on the Map of Varanasi, Shiva's Sacred City. *Note*: Original drawing by Gutschow (2006), Tracing and editing by the author

Group 2: Vāpī (a well with steps): (4) Jñāna Vāpī (5) Karoṭaka Vāpī.

Group 3: Kūpa (a round well without steps): (6) Kardama Kūpa, (7) Oṁkāra Kūpa, (8) Mahākāla kūpa, (9) Vṛddhakāla Kūpa, (10) Kāmākṣī Kūpa.

Group 4: Tīrtha (a sacred stream, river, place in the vicinity of sacred spring): (11) Kardama Tīrtha, (12) Puṣkara Tīrtha, (13) Sannihita Tīrtha, (14) Shaṅkhūdhārā Tīrtha, (15) Cakrapuṣkariṇī Tīrtha, (16) Ratnacūḍāmaṇi Tirtha, (17) Vaitarani Tirtha, (18) Mandākinī Tirtha (see Figuare 2.8), (19) Vimala Tirtha.

Group 5: Kuṇḍa (a water reservoir with oblong shape): (20) Hanumadi Kuṇḍa, (21) Durgā Kuṇḍa, (22) Reṇukā Kuṇḍa, (23) Kṛmi Kuṇḍa, (24) ...Kuṇḍa (no identification), (25) Vāsuki Kuṇḍa, (26) Bakariyā Kuṇḍa, (27) Gaurī Kuṇḍa, (28) Revā Kuṇḍa, (29) Lakṣmī Kuṇḍa, (30) Kāmākṣī Kuṇḍa, (31) Lavakuśa Kuṇḍa, (32) Dadhavi Kuṇḍa, (33) Sūrya Kuṇḍa, (34) Pitṛ Kuṇḍa, (35) ...tra Kuṇḍa (not identified), (36) Mātṛ Kuṇḍa, (37) Vārāhī Kuṇḍa.

Group 6: Sarōvara (a lake or a large pond): (38) Yūpa Sarovara, (39) Māna Sarovara.

Figure 2.8: Benares from the Mundakinee Talao by James Prinsep, 1831 The British Library Board, London. *Note*: Mandākinī Tīrtha

Group 7: Sāgara (a sea, or a pond): (40) Gandharva Sāgara, (41) Tata Sāgara.

Group 8: Untitled: (42) Pṛthvīśvara, (43) Kāpiladhārā, (44) Lolārka, (45) Kurukṣetra, (46) Ṛṇamocana, (47) Lāṭ Bhairava,[10] (48) Pāpamocana, (49) Kapālmocana, (50) Tarni (uncertain), (51) Īśvaragaṅgī, (52) Ghaṇṭākarṇa.

Group 9: No description: (53) (blue rectangular tank), (54-55) (red rectangular tank), (56) (small rectangular tank).

The mapmaker has drawn the shape and other architectural forms as described in the Purāṇas and other literary texts. The nine categories represent the diverse forms of water bodies and their locations respective to other water bodies and temples (see Figure 2.9).

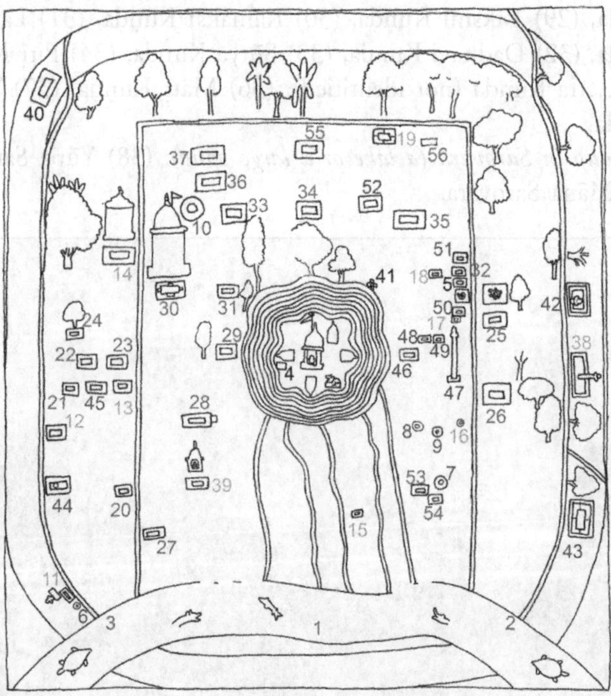

Figure 2.9: Tracing of all the sacred water bodies (including the two rivulets and a few other associated forms) on the 'Map of Varanasi, Shiva's Sacred City'. *Note*: Drawing by the author

WATER SENSITIVE MAPPING

The eighteenth-century pictorial map of Varanasi city depicts 183 temples (Gutschow 2006) and 56 distinct types of water structures. These water structures include the principal river, rivulets, wells, lakes, ponds, and water tanks of different shapes. The water tanks mentioned in Group 8 are subtitled as ponds, water reservoirs, tanks, wells, etc., in other works of literature. Three water structures under Group 9 have no description (Gengnagel 2011).

Gengnagel (2011) argued that the mapmaker, most likely a visitor and not a resident of Varanasi, had not mapped the lofty and striking riverbank and nearly failed to portray the spatial understanding of the city.

However, the pictorial map of Varanasi explained effectively the geomorphological features of the city with one principal river flowing from south to north, tributaries, small streams and gathering of inland water tanks and ponds. The mapmaker captured the modest slope of the tract from west to east direction and probably oriented the map to the westward direction (Joshi 1965: 4). The map also subtly depicted the natural flow of water and the fluid nature that dominated the landscape of the city. It successfully captured the movements of the water and highlighted the catchment ecology. The pictorial representation of the city arguably depicted that land and water are not separate but interwoven without fixed demarcation lines.

CONCLUSION

According to Kwa Chong Guan (1992: 15), the cartographer created a visual presentation that was not necessarily confined to two-axis but how the physical environment around it is felt. Eck (1998), elucidated the potential of 'imagined landscape' constructed by mapmakers which captured more attention than the Western scientific map.

James Prinsep's map of 1822 was a topographical map; a map that essentially depicted the physical milieu (Birch 1949: 1). Even though, Prinsep had shown the location of temples and marked the dotted line to depict the Antargṛha Kṣetra (innermost circuit) the central feature of this map was an empirical representation of the colonial city. Prinsep's map does not require any supplementary knowledge or context to see the map. The first-time visitor to Varanasi city can effortlessly locate the places on this two-centuries-old map after a careful glance. However, in the case of the pictorial map of Varanasi, it is not intended as a guide map. Instead, it is an assemblage of all celestial edificed that harmonizes with the surrounding natural landscape of the city. To read the pictorial map of Varanasi, one needs to read the supplementary texts. It is a dynamic representation that subtly blended the physical and spiritual forms which both cartographers and devotees are aware of. The mapmaker translated the spatial texts that discussed the geography of Varanasi into the pictorial format.

The two distinct maps discussed in this chapter, juxtapose the fundamental perception of the land and water ecology and their interconnection. The pictorial map of Varanasi with its westward orientation hints at the natural slope and a high natural embankment on the River Ganga. The peculiar arrangement of the various sacred zones and the water pools point out that the mapmaker was probably aware of the Purāṇas and other literary texts that provide detailed spatial information about the region. He therefore possibly oriented the map westward to depict the gradual gradient and the drainage pattern of the city. The map reveals the continuum between land and water. The careful observation enables one to see the flow pattern of the city and epitomizes the notion that water is omnipresent in Varanasi. The mapmaker of the pictorial map subtly hinted at the sacred landscape of Varanasi which is not static but dominated by flowing water.

Prinsep's map of 1822, is an archetype of Western mapping where scale, dimensions, orientation and careful survey are vital benchmarks that provide the foundation to produce an accurate representation of the city. It is an objective representation of the urban centre with an intention for colonial administration and control. The map drawn in 1822 reveals the trajectory of the survey of land and revenue system with its underpinnings of land as a 'fixed' and property notion while water as a mere resource. The 1822 map distinguished between the land and water and shows the rigid boundaries of these two elements.

NOTES

1. For more information about this map see: Niels Gutschow, 2006, *Benares: The Sacred Landscape of Varanasi* and Jörg Gengnagel, 2011, *Visualized Texts: Sacred Spaces, Spatial Texts and the Religious Cartography of Banaras.*
2. According to the 'object history note' on the website of Victoria and Albert Museum, London, the nineteenth-century museum label described this as 'Pictorial Chart of Shivapura'. http://collections.vam.ac.uk/item/O434358/painting/
3. Kāśī Khaṇḍa tells the story of King Malayketu (Malyaketu) offering a 'picture' (*citrapaṭa*) to Kalavati. After seeing the picture painted on a canvas, Kalavati, pointed out various sacred locations on the canvas and told the name of the place and the prominent temple or river near that place. From this narration, it is quite possible that the picture was a map-like projection of Varanasi. (*SP* KKh. 33.96-180).
4. Maya Prasad Tripathi (1969) referred to *Śatapatha Brāmaṇa*, Purāṇa-s, etc., to discuss how these texts widely mentioned survey methods and cartography practices. He enumerated a number of Sanskrit words such as *lekhya, parilekha, samputakā* corresponding to English words such as map, cartography and atlas.
5. Even though the production of the map was his interest for the city, Prinsep received Rs. 1,000 as remuneration from the British government.
6. *Nālā* is translated as a small stream. However, the post-1822 colonial maps labelled Varana as a river.
7. Mapped and identified by the author.
8. River Varana is considered as Pingalā (right tubular vessel), and Asi rivulet is considered as Iḍā (left tubular vessel). The territory in between these two rivers is called Avimukta kṣētra. (*SP* KKh. 5.25).
9. The five streams are called Pañcanada (Pañca-Ganga) (five rivers: Ganga, Yamuna, Sarasvati, Kiraṇā, and Dhūtapāpā,). These rivers are a symbolic representation of rivers that appears in Purāṇas.
10. There is no visual representation of a tank near the pillar. This has been added because the Lāṭ Bhairava (Daṇḍapāṇi Bhairava Tīrtha) temple tanks mentioned in Kāśī Khaṇḍa of *Skanda Purāṇa.*

CHAPTER 3

Draining the Wet City

The Victorian sanitary measures, reform policies and its arbitrary implementations had tampered largely with the natural settings of Varanasi. As a result of this, water bodies were either eliminated or turned into stagnant and non-functional tanks. This chapter thus focuses on the colonial annexation of the former state of Benares (Varanasi) followed by peremptory implementation of the urban reforms and development initiatives. However, it is essential to note that the conception and implementation of these 'reforms' of the city were carefully calibrated with the Indian debt, fiscal restraints and strategic importance of the city and the presence of European residents and soldiers.[1] The chapter, therefore, draws upon the significant events in Britain and India during the nineteenth century that shaped the urban reforms policies considerably. In this chapter, it is argued that these events, essentially transformed the colonial priorities and introduced modern systems to mitigate the risk and get better control of the territory and the subjects.

The nineteenth century of Varanasi presents crucial evidence to track the elimination of sacred water pools. In this chapter, it is demonstrated how the early modern sewage works affected the waterscape of the city greatly and triggered the pollution of the river and disruptions in the connectivity and natural flow of the water. The modern term of 'reclamation' of these water pools which were first labelled as marshes and swamps and in time turned into gardens, was exercised to validate the work.

Anant Altekar, in his book (1947), explored the role and contribution of Marathas and how they reconstructed Varanasi. He succinctly summarized 'Modern Benares is largely a creation of the Marathas' (Altekar 1947: 24). However, in the contemporary context, this is instead a very restricted assessment of the city. In this chapter, it is argued that contemporary Varanasi is mostly an outcome of ad hoc colonial urban policies and their disorganized implementation shaped by events that occurred in nineteenth century Britain.

WHEN WATER WAS CLEAR AND ABUNDANT

Scholar and traveller Hiuen Tsiang (who visited Varanasi during the seventh century) described a densely populated city Varanasi as 'Po-Lo-Ni-Sse' placed on the west side of River Ganga (Beal 1884: 44-5). He referred to the principal deity as 'Ta-tseu-tsai' (Śiva or Maheśvara) and while illustrating the natural setting of the city wrote about the 'pure streams of water' flowing around the numerous temples. This account is probably the first accessible foreign record that mentioned the pure streams of water categorically. A review of annals reveals that Varanasi was visited by many foreign travellers and seekers such as, al-Biruni (1021-5) who visited Varanasi which he described as a revered place closely associated with Hindu law and religion. During his travel, he also encountered one of the sacred water tanks (*kuṇḍa*). He had an interesting account where he described the tradition of building a water tank for 'ablution' at the consecrated site. He recounted the structural form, material that was used to build the tank and the general spaciousness as a feature of the entire site. He wrote,

> In every place to which some particular holiness is ascribed, the Hindus construct ponds intended for the ablutions. In this they have attained to a very high degree of art, so that our people (the Muslims), when they see them, wonder at them, and are unable to describe them, much less to construct anything like them. They build them of great stones of an enormous bulk, joined to each other by sharp and strong cramp-irons, in the form of steps (or terraces) like so many ledges; and these terraces run all around the pond, reaching to a height of more than a man's stature. On the surface of the stones between two terraces they construct staircases rising like pinnacles. Thus (,) the first steps or terraces are like roads (leading around the pond), and the pinnacles are steps (leading up and down). If ever so many people descend to the pond whilst others ascend,

they do not meet each other, and the road is never blocked up, because there are so many terraces, and the ascending person can always turn aside to another terrace than that on which the descending people go. By this arrangement all troublesome thronging is avoided.[2]

These early mediaeval accounts help one to reckon on the architectural aspects and functions of tank building and assess the prominence of these water tanks in social settings.

After the late sixteenth century, records reveal the arrival of European and English travellers visiting and narrating the detailed account of their experience in Varanasi. Ralph Fitch[3] is considered one of the earliest English travellers arriving in India in the sixteenth century. Fitch sailed down the Yamuna River and after visiting Prayag (1583-91), he reached Varanasi. He described Varanasi as a 'great town' of Gentiles and witnessed the greatest 'idolaters' (Ryley 1899: 103). A well-known traveller, writer and doctor, Niccolao Manucci (1638/9-1717) also visited Varanasi (during the seventeenth century) and referred to a saying *tora cana banarismo Rana* meaning, little to eat, but reside in Banaras. William Irvine (1907: 83) who translated Manucci's travel account titled *Storia Do Mogor* explained it by adding that Varanasi is a 'nice place, with a good climate, productive land and cheap food'. Irvine recounted his experience and elucidated Benares as 'Bana-Ras; Perfect Delight'. It was found that almost similar experiences echoed in the books of travelogues written by Russian and Georgian travellers visiting Varanasi in the late eighteenth and early nineteenth century. A Georgian traveller named Rafail Danibegov (Danibegashvili) who stayed in Varanasi found the city magnificent and healthier than other cities located on the River Ganga.[4]

EARLY EUROPEAN VOYAGES TO VARANASI

After Ralph Fitch, there were numerous travelogues, diaries and anecdotes of European merchants, physicians, travellers, officers, surveyors and missionaries who wrote on the urban milieu, prevailing customs and worship practices, and showed interest in the history and geography of the second-largest city in the 'United

Provinces', during the colonial regime. The initial surveys and understanding of the Varanasi region depict conflicting observations[5] made by colonial officers and how the perception gradually changed owing to the eruption of certain events and calamities. The reading of these accounts and the pencil and colour drawings compellingly demonstrate how the perception and notions were transformed as the traders and travellers became the rulers.

Before revisiting the colonial narrative, one can take a look at an interesting account of Dean Mahomet,[6] who was originally from Patna, India. He wrote a series of letters to his friend in English. In one of the letters, he wrote about Varanasi when the region was partly under the dominance of the East India Company. Along with the detailed description of the city, he described the water tanks.

> In different parts of the town, there are tanks, or well for the use of the citizen and all the refreshment of passing strangers, who if in indigence, are also humanely supplied with food by persons employed to attend at the tanks for this very purpose.[7]

Figure 3.1: 'Dusasumade Gaut, at Benares on the Ganges' by Thomas Daniell RA, 1795. The Victoria and Albert Museum, London
Note: Daśāśvamedh Ghāṭa on River Ganga, Varanasi

Dean Mahomet's brief description of the water tanks shed light on the usage of the public water tanks punctuated in various parts of the city. This description highlighted the functionality of water tanks and their wider utility.[8] During the middle of the nineteenth century, Reverend William Buyers travelled from Calcutta to Agra and later to Delhi. Reverend Buyers on his road journey to 'pilgrimage place' Varanasi saw many masonry wells and large-sized water tanks built mostly by wealthy residents (Buyers 1848: 225). These community tanks and wells epitomized the art of building water harvesting structures and their diverse usages.[9]

A VICTORIAN PROGRESS EPOCH

The late eighteenth century marked the disintegration and fall of Muslim rulers and the rise of the Maratha powers regaining the territories, including a limited control of the Varanasi region (Medhasananda 2002: 16). During their brief control over the region, many Maratha rulers patronized the reconstruction of many temples, *ghāṭas* (Eck 2015: 90) and temple tanks of the city. After a transitory succession of Kashi Raj and following a brief period of turmoil eventually in 1794 'Benares state' was entirely annexed by the British government.[10]

Jonathan Duncan was 'Resident at Benares' from 1787 to 1795. Even though the jurisdiction of the Resident was undefined (Shakespear 1873: iv), the actual authority was vested in Duncan (Cohn 1960: 424). During his tenure, Duncan allocated surplus revenues of the province for building the bridge on the Varana River, which would connect the cantonment and the old city of Varanasi.[11] Before the annexation of the state, the sanitary measures were supported mainly by the annual contribution fund created by the local merchants.[12] The correspondence during late eighteenth century revealed that requests were often made by wealthy merchants, traders and residents to renovate and cleaning of the water tanks that had either dried up or deteriorated.[13]

However, the Duncan administration blamed the narrow and crowded lanes of the city and stagnant drains which were polluting the 'air in the heart of the city' and endangering the health of inhabitants (Oldham 1876: 208). As one of the improvement measures a trench was dug to carry the filthy water in the Varana stream in the northern part of the city. However, this exercise did not yield any success. Duncan was unsatisfied with the state of sanitation of the city and constructed public latrines and installed waste containers using the fees and fines collected by the court (Naranin 1959: 182).

In the late 1820s, James Prinsep arrived in Varanasi as the Assay Master at the Benares Mint. Besides his duties as an Assay Master, Prinsep made a significant contribution as a scholar, epigraphist, archaeologist and left behind the extraordinary series of drawings of the city and its landscape.

In his initial period along with preparing a map, Prinsep also conducted a detailed survey of houses and census of the city. Before the census of Prinsep, Zulfiqar Ali Khan, Kotwal of the city was ordered by Deane, the resident of the city to count the number of houses and provide the estimate of the population (Prinsep 1832). Zulfiqar Ali Khan, based on the number of storeys of the houses estimated 582,625 inhabitants (Valentia 1811) of the city. Prinsep rejected this estimate stating that the numbers were overstated. Prinsep's exercise of a 'correct census' (Prinsep 1832: 480) facilitated the effective implementation of various taxes.[14] Matthew Edney (1997: 296) elucidated how Europe's centralized states realized the potency of information of the dominions for better authority and expanded the realm of the tax revenues. During his ten years of tenure, Prinsep also took an interest in various sanitary reform projects in Varanasi (Sherring 1868: 24, 114-15). Prinsep built the first sewer system replacing the 'Nawabi Drain' in 1829.[15]

Reverend Sherring (1868: 114-15) who described the draining of 'Matsyodarī Tālāba' (pond) by James Prinsep[16] (see Figure 3.2) and justified the complete elimination of this extensive pond on sanitary grounds and admired it as a 'meaningful' measure.[17] The drying up of the Matsyodarī Tīrtha resulted in the decline in devotees visiting the temple of Kāmeśvara[18] located on the eastern banks of the Matsyodarī Tālāba. Sherring referred to the temple of Kāmeśvara as 'Kamananath' which was closely connected with the Matsyodarī Tālāba and added that earlier devotees used to bathe

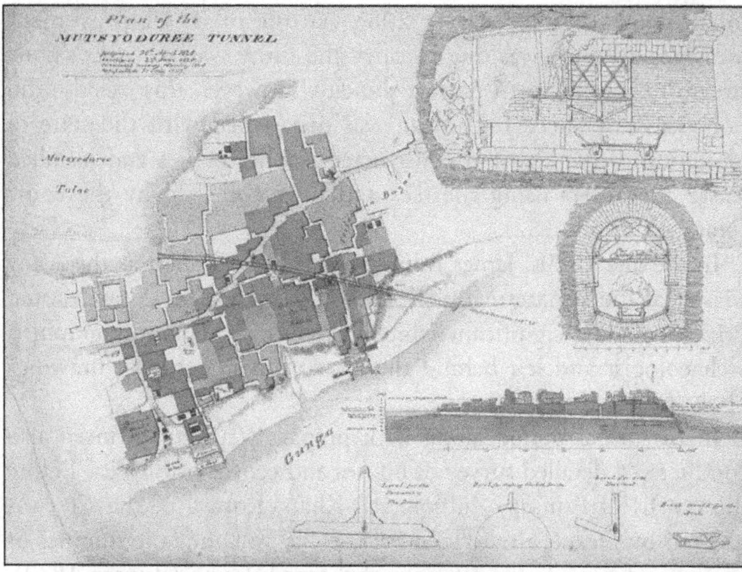

Figure 3.2: Map and cross-section of the drainage tunnel from the Machhodri (Matsyodarī) tank to the river, Benares, with elevation and cross-section of the tunnelling process and drawings of the equipment used by James Prinsep, 1825, The British Library Board, London

in the pond and then proceed to the temple to pay homage to Kāmeśvara. Prinsep's map of 1822 described the extensive size of the Matsyodarī Tālāba and the location of Kāmeśvara Temple.

Sherring's account of the decline of the prominent *tīrtha* underlined the active linkage between the temple, water pool, and the devotees. The removal of the sacred water tank disrupted the link that probably sustained the temple and the water reservoir. Singh (1994: 226) defined the temple tanks and water pools as 'liminal places'. Water sites are spatial passages functioning as an intermediate between physical and celestial spheres.[19] Singh argued that water tanks are more prone to obscurity and hence might lose their distinctiveness in the physical settings in due course of time. However, he added that these places would preserve their prominence.

On one of the other detailed illustrations of the Matsyodarī Tālāba[20] and a proposed drain, (see Figure 3.3) James Prinsep explained the channel drawn by him connecting the Hālū Gaḍḍhā,

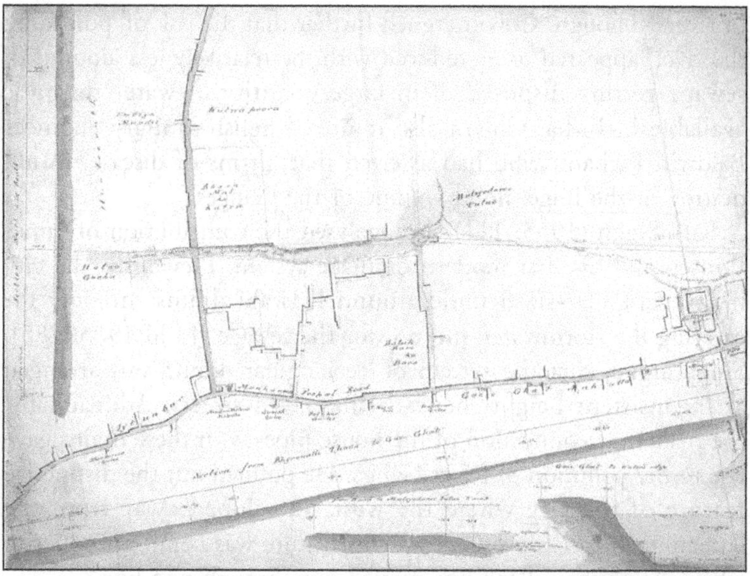

Figure 3.3: Map of the area around the Machhodri (Matsyodarī) tank, Benares, with cross-sections showing relative water levels of the marsh and the river, and the site of the proposed drain. With notes and measurements by James Prinsep, 1825, The British Library Board, London

and Matsyodarī Tālāba calling it the 'Channel stopped up by rubbish and filth'. The construction of new streets and apathy towards the maintenance turned the water tanks and streams into receptacles of silt and filth.

H.R. Nevill (1909: 264) reviewed the drainage work of the city and criticized the 'unscientific system of rectangular drains' constructed by James Prinsep[21] in 1823 severely. He explained the rudimentary nature of the construction that triggered the contamination of river water and polluted the city. Nevill also discussed how often, due to the financial restraints, many construction projects were curtailed. Interestingly, Sir Auckland Colvin (1894: 522) who was the Lieutenant Governor of the North-Western Provinces and Chief Commissioner of Oudh, in his paper titled 'Municipal and Village Water-Supply and Sanitation in the Northern West Provinces and Oudh', revisited the model of establishing 'sewage farm' and abandoning the disposal of sewage in rivers. Though, Colvin argued further that the risk of 'poisoning the river' appeared to be reduced with the relatively less amount of sewage getting disposed of in large quantity of water that was available in India. Colvin also quoted English sanitary engineer Baldwin Latham, who had asserted that 'germs of disease' would destroy in the huge 'flood volume of the Ganges'.

R.L. Singh (1955: 111) also analysed the contribution of James Prinsep and his first modern drainage system. Previously, the city had rectangular-sized underground Nawabi drains[22] mainly for carrying the stormwater and not for the sewage (Joshi 1965: 285). Singh argues that the system of 'rectangular drains' was arranged at inconsistent heights beneath the paved streets. Additionally, the makeshift connection of the house pipes with these drains gave rise to the pollution of River Ganga. He pointed out the distinctive feature of the city where the west side sloped away from the embankment on River Ganga. This feature was counterproductive when the entire sullage was carried out through one line.

This 'combined system' of carrying the stormwater and sewage together in one huge tunnel was a standard process in England and thus also implemented in India till the late nineteenth century (Richardson 1887). However, Edwin Chadwick[23] (1800-1890) one of the leading architects of the 1834 Poor Law Commissioners, when consulted recommended the 'separate system' for the drainage plan of Cawnpore (now Kanpur). Chadwick's plan suggested that stormwater would be carried to the river and sewage through a separate pipe to the farms. Veena Talwar Oldenburg (1984: 143) in her brilliant investigation of Lucknow city during the colonial regime pointed out how Chadwick's suggestions and understanding of the broader sanitary realm of the urban areas were not comprehended by the British authorities posted in India. The majority of the funds and resources were spent on the civil stations and in cantonments, while the old city areas and outskirts were treated either as marginal or simply overlooked.

James Prinsep constructed another drain to dry out one another prominent *tīrtha* called as Mandākinī Tālāba (see Figure 3.4). Prinsep (1831) described this water body as part of a series of ponds that were running parallel to River Ganga. Earlier this backchannel

Figure 3.4: 'Distant view of Benares with two men catching turtles in the foreground from the Mandakini tank' (in watercolour) by James Prinsep, 1826, The British Library Board, London. *Note*: Mandākinī Tālāba

used to contain the excess floodwater of the river during the downpour. Prinsep also added an interesting episode of transporting more than 1,500 turtles from their original habitat of Mandākinī Tālāba to River Ganga as the site after drying up was converted into the grain market. Prinsep continued his description and reflected on the adoration of turtles by Hindus who also used to feed them.[24] The number of turtles indicated the spread of the water reservoir and its functional nature.

Due to defective sewage systems and negligence, sacred water pools and streams such as Misir Pokharā (Miśraka Tīrtha) and Godauliyā Nālā (Gautama Kuṇḍa, Gōdāvarī Tīrtha) were drained (Singh 1955: 88). Nevill (1909: 265) described the 'reclamation' work of Misir Pokharā 'marsh' and Godauliyā Nālā. He admired the work by calling it the 'first attempt at drainage on scientific lines'. After the elimination of the large-sized temple tanks and ponds, a few tanks were dug as a symbol for the eliminated tanks. They later became a subsidiary part of public gardens.

In 1823, a local improvement committee was formed to undertake the preparation for the public works in the city (Medhasananda 2002: 429-33). James Prinsep, with his active interest and experience, was appointed as a secretary. The appointed committee suggested various reforms and renovation of the existing infrastructures and edifices. However, some of the projects were rejected by the government. Finally, in 1829, the local committee was dissolved by stating fiscal limitations. Some of the projects like establishing a marketplace on the reclaimed land were financed by James Prinsep due to the winding up of the committee.

THE EMERGENCE OF SANITARY MEASURES IN LONDON

To understand the core of colonial sanitary measures in Varanasi, one needs to revisit the city of London and what happened in Victorian cities during the nineteenth century. The urban population of Britain saw enormous growth between 1801-1901 (Ward 1994: 11), giving rise to an increase in new urban cities (Coleman 2012: 1). The demographic change also brought new challenges such as overcrowded localities, poor standard of housing and an increase in pollution and waste (Harrison 1990: 15-16). During the nineteenth century, the city of London and its residents were concerned with the filthy streets, horse dung, waste adding more menace and smoky air. Jackson Lee (2014) described the series of Victorian reforms to tackle the various problems of London. Lee spoke about two important things about the Victorian capital that closely reflected the civic reforms that were initiated in colonial Varanasi. The first thing was about the prime function of London's city sewers at the beginning of the nineteenth century. Sewers were meant to carry out the stormwater and not the excreta. The second thing was about the non-availability of public urinals for 'men' till the mid of nineteenth century, causing a nuisance.

This period also saw the frequent outbreak of epidemics such as influenza, cholera and typhus. The physicians and researchers thought that marshy land, stagnant water and accumulated waste caused the 'miasma' or effluvia.[25] Edwin Chadwick and Doctor

Thomas Southwood Smith were assured and vehemently argued about the miasma theory and its connections with the outburst of diseases. The 'Report on the Sanitary Condition of the Labouring Population of Great Britain' of 1842 comprised the multiple surveys and findings of medical professionals corroborated with detailed maps and raised concerns for public health and welfare. The report repeatedly confirmed that swampy areas, open cesspools, stagnant ditches, decaying animals caused the toxic odour. The reviewers saw a commonality of narrow and dirty streets, open drains, filth, dunghills, swampy areas and wet surfaces around all the working-class localities.

As a brainchild of Poor Law Commissioners Chadwick alleged that the epidemics were caused because of the 'atmospheric impurities' generated by the decaying animals and vegetation, piles of garbage and filth, uncovered sewers, overcrowded dwellings, marshy areas and lack of proper ventilation. The Report categorically asserted that the prevailing law to safeguard public health was inefficacious. Additionally, the report advocated the need for the improvements in public administration, water supply and the setting up of better sewers and drain to avoid river pollution. In the last part of the sanitary report, Chadwick stressed the total cleanliness at all levels to improve the 'moral condition' of the people.

MAPS, DIAGRAMS AND NUMBERS

The detailed maps of towns like Leeds and other areas in London were prepared to designate the localities where cholera and other infectious diseases were found. The report also contained the architectural drawings of the cottages and houses from different counties to illustrate the ventilation and light arrangements for the improvement of health and comfort of labouring people. The report underlined the importance of 'studious planning'. The report was substantiated with the enormous amount of data, census reports, statistical tables,[26] maps, plans, diagrams and revealed the obsession for quantification. The report also discussed the poor condition of water supply and water charges and criticized the role of private water supply in towns and rural districts of England. The underpinning notions, tools and technologies were later reflected in the sanitary measures and urban reforms of Varanasi.

The sanitary report projected many theories about the possible reasons behind epidemics like cholera, malaria and typhus. It was found that along with the overall unhygienic surroundings and crammed localities, poor ventilation, wet areas and 'humid weather' also deteriorated the condition. These findings were crucial as many concerns surfaced in the report and the discourse influenced the perception of British authorities. David Arnold (1993: 33-4) substituted miasma with 'malaria'. He explained that experts in colonial India were incompetent to determine the root cause[27] of the epidemics, but, the unhygienic localities were certainly breeding ground for the disease. Health experts supposed that cholera[28] spread in 'tropical'[29] India through the toxic odour from decaying vegetables, congested localities and body waste. Interestingly during 1854, when the miasma theory with its roots in the 'Greek and Roman medicine' was gaining ground, John Snow examined the cholera outbreak cases in some of the areas of London city. Snow claimed that the disease spread due to contaminated water sources and thus paved the way for the 'germ theory'.[30]

Contemporary to the period when sanitary reforms and public health concerns were getting momentum in Victorian cities, the extensive 'Report from the Select Committee on the Affairs of the East India Company' was tabled before the House of Commons in 1832.[31] This report critically analysed and commented on the history, religion, law, political institutions, customs, worshipping practices and social norms of the people. It revealed how the Europeans portrayed the heathen worshipping practices and described the deities and surroundings of the temples. The report of 1832 stated:

> Besides the consequences of idolatry, which are universal, the very appearance of the Hindoo idols, in general, has a tendency to degrade the worshippers: they are hideously ugly, with many heads, arms and weapons; with great teeth and eyes, and terrific countenances, of a black colour; many of them smeared with oil, and smelling strongly of it. They are shut up in narrow dark rooms, and the approach to them is through obscurity and silence. The scandalous obscenity of

others has been before hinted (at). Figures of the monkey, the bull, and of various other animals, are often placed around and worshipped as the attendants of the superior deity of the place.[32]

ENLIGHTENMENT OF NATIVES BY MISSIONARIES

Varanasi was amongst one of the earliest provinces annexed by the East India Company[33] and thus enabled a passage for the missionaries to commence the teachings of the Gospels. The first missionary work[34] started in Varanasi was in 1816 (Kennedy 1884: 21). James Kennedy was one of the earlier missionaries who stayed and worked in Varanasi province for a long time and witnessed some of the critical events, portrayed the Europeans and native population vividly. Kennedy writes that as a missionary, he was always treated and remained as an outsider amongst the natives. However, he found easy access in all European establishments. For Kennedy Varanasi was a commercial as well as a religious city where innocent and ignorant people and pilgrims were exploited by the priests in the hideous temple and suppressed by superstitious practices (Kennedy 1884: 69-87). Sherring (1884: 175), in his work on the history of Protestant missions in India, offered a solution for the ignorant and enslaved natives. He wrote:

> The religion of idolatry, of sculptures, of sacred wells and rivers, of gross fetishism, of mythological representations, of many-handed, or many-headed, or many bodied deities, is losing in their eyes its religious romance. English education based on the Bible has produced a revolution of thought in their minds.

Both Kennedy and his friend Sherring were confident enough that Varanasi as a 'principal seat' of the Hindu philosophy was drying up at its 'fountainhead'. They believed that Western education and science would transform and gradually, Hindu men will change their habits and lives (Sherring 1868: 358). The missionaries perceived that Hindus would discard the fetishism and the temples which were repulsive. Christianity and Christian nations were posed as antipodal to Hindu practices, which were not only civilized and the land of enlightened people, but all great inventions such as steam power, railways, the telegraph, etc., were invented in these progressive nations (Murdoch 1897: 37-8).

Sherring's account reflected the concern of the European mind about the public health of natives and the overall sanitation of the ancient city. He criticized that during the regime of previous Muslim rulers, no significant progress had taken place. Sherring also discussed the miasma theory and linked it with the 'foul wells and tanks' of the city. One can find many points of resemblance and terms that were expressed in Edwin Chadwick's 1842 Sanitary report. Sherring (1868: 181-2) described the condition of the tanks as:

> But, when we reflect on the foul wells and tanks in some parts of the city, whose water is of deadly influence, and the vapour from which fills the air with fever-fraught and cholera-breeding miasma; when we consider the loathsome and disgusting state of the popular temples, owing to the rapid decomposition of the offerings, from the intense heat of the sun; when we call to mind the filthy condition of nearly all the by-streets, due to stagnant cesspools, accumulated refuse, and dead bodies of animals; and when, in addition, we remember how utterly regardless of these matters, and incompetent to correct them, is the police force scattered over the city, the difficulty becomes almost overwhelming.

Many Western authors during the colonial period, attributed the term 'sacred' or 'holy' to the water tanks and rivers in India in their writings. However, they were confused and confronted with the 'Eastern'[35] notion of 'cleansing the sin' by merely bathing or plunging into the *kuṇḍa* or in the river, mainly when the river or the tank was filled with silt and mud (Greaves 1909: 49-50).

Arnold (1993: 186) explained how 'missionary propaganda' had dominated the colonial dispensation and constructed the narrative to establish a link between the Hindu pilgrimage places and diseases.

HINDU PILGRIMAGE PLACES AND DISEASES

In the early and middle part of the nineteenth century, the cholera epidemic caused massive death rates in Europe and North America. An English writer and traveller, Isabel Burton (1879: 86) discussed

Figure 3.5: The Monkey Temple, Benares. Photograph by Samuel Bourne (1863-6). *Note*: Durga Temple and temple tank of Varanasi. The Rijksmuseum Amsterdam Collection

the spreading of cholera. In her book, she listed 'nine conclusions' of the pandemic. She identified India as the 'special genesis' of cholera and asserted additionally that it was India—from where the disease spread over the 'civilized world'. The origin of frequent outbreaks of cholera epidemic was 'located' in the Hindu pilgrimage sites of India in the International Sanitary Conference convened in Constantinople in 1866 (Harrison 2011: 290). During the auspicious days of Kumbha Melās, thousands of pilgrims bathed in the sacred rivers and tanks and carried the water back to their homes. Thus, the water of pilgrimage places and mass congregation both were identified as sources of spreading water borne diseases (Arnold 1993: 184). Since pilgrims visited around the year, Varanasi was considered as most responsible for the epidemics (Fisher et al. 1884: 99). The successive fourth conference of 1874 in Vienna, studied the conclusions of the Constantinople Conference of 1866 and reaffirmed that cholera was exported to other parts of the world from its origin in the valleys of River Ganga and its pilgrimage places in India (Howard 1975: 30). The British rule was uneasy to accept this but implicitly acknowledged it by stating the 'sanitary improvements' measures at many portcities in India and also to prohibit certain unhealthy customs of Hindus.

REFORMS AFTER THE 'WAR OF 1857'

European settlements owing to many reasons such as safety and the notion of superiority, were segregated from the core city area of Varanasi. In the initial phase, the Company administration started constructing the courts, hospitals, stores, and residential bungalows for European officers in Sikraul (Secrole); the north-west part of the city. Prinsep's map of 1822, displayed the settlement of the Sikraul and Cantonment area separately. The British accounts cautiously controlled the narrative of the 'War of 1857' and preferred to constrain it to 'sepoy mutiny'. Both Sherring and Kennedy narrated the events from late 1856 and what happened particularly after 4 June 1857 when the native infantry regiment stationed in Varanasi took up the arms and joined the war. The situation of the city remained volatile for several days. However, the city remained comparatively peaceful and under control.[36] The War of 1857 marked the closure of the role of the East India Company in India and powers were transferred to the British government under the Crown. Kennedy (1884: 213) recalled that after the war, the strength of the English army and officers from public works increased in Varanasi. The terrible events of the war compelled the colonial regime to revise the overall safety and maintain the health of their soldiers and other European residents.

After the War of 1857, in 1864 the commissioners submitted their voluminous report of an inquiry into the sanitary state of the Queen's army serving in India.[37] The commissioners went through the statistical reports and sanitary documents of India and claimed to cover the extensive and diverse territory and corroborated with the shreds of evidence. The Report located the vast floodplains of the Ganga River as the epicentre of diseases such as cholera. The report affirmed that cholera, malaria and other epidemics could quickly spread in the 'imperfect natural drainage', moist and low-lying areas, stagnant water tanks, marshy grounds, etc.

TABLE 3.1: DETAILS OF BRITISH NORTH-WEST PROVINCES IN 1852-3

Districts	Square	Population	Density
Agra	9,298	4,373,156	465
Allahabad (Prayag)	11,971	4,526,607	378
Benares (Varanasi)	*19,737*	*9,437,270*	*478*
Delhi	8,633	2,195,180	254
Meerut	9,985	4,522,165	453
Rohilcund (Rohilkhand)	12,428	5,217,507	419
Total	72,052	30,271,885	420

The above census data of Varanasi as one of the major cities from the North-West Provinces demonstrated the most extensive area with almost twofold population size compared to other significant power and trading centres.

J. Ranald Martin, a member of the Royal Commission on the Sanitary State of the Army (1863) had discussed the health of the British army in India (Hume 1986: 708). Martin underlined the paramount importance of the health of British soldiers in the tropical climate. By comparing the ground data, Martin compared the number of healthy and fit soldiers and pointed out the significant number of ailing men in the hospital. According to Martin, the life of one soldier with his arms cost roughly 100 pounds for Britain and thus every sick military personnel was a substantial financial burden on Britain.

The average strength of soldiers in Varanasi was the second largest after the Lucknow station. The number of daily ailing soldiers of Varanasi was amongst the highest figures including Delhi and Dinapore (now Danapur) stations. In his book, Martin also discussed the issues stated in the Royal Commission enquiry and offered brief information on the deplorable condition of native towns. He (Martin 1868: 39) wrote:

The towns and bazaars in the vicinity of lines were in the worst possible sanitary state—undrained, unpaved, badly cleansed, often teeming with offensive and dangerous nuisances—with tanks, pools, and badly-made surface gutters, containing filth and foul water; the area overcrowded with houses, put up without order or regularity; the external ventilation obstructed, and the houses overcrowded with people; no public latrines, and every spare plot of ground covered with filth in consequence; no water-supply, except what was obtained from bad shallow wells and unwholesome or doubtful tanks. These towns and bazaars were the earliest seats of epidemics, especially of cholera, before their ravages extended to the European troops in the vicinity.

TABLE 3.2: MORTALITY IN HOSPITALS
(FIGURES OF VARANASI WERE EXCERPTED)

Station	Average strength	Ratio per thousand of average		
		Daily sick	Admitted	Died from cholera
Benares (Varanasi)	1,081	75.7	1,692	5.6

Source: Excerpt from Bengal Sanitary Commission. First Annual Report of the Sanitary Commission for Bengal, 1864-5, p. 7.

Multiple sanitary reports since the late eighteenth century and during the nineteenth century squarely blamed the stagnant water tanks and directly linked them with the epidemics.[38] Arnold (1989: 248) highlighted the penetration of colonial power and its dominant discourse of modernity forced the sacred water and its healing qualities declared profane and polluted. He further argued that this Western perspective cleared the path of forming a municipal board, which would make city life healthy and clean.

TOWARDS THE MODERN DRY CITY

Thousands of pilgrims used to visit sacred places such as Varanasi, despite the lack of transportation facilities and additionally, paying the pilgrimage tax to the rulers and later to the East India Company.[39] Swami Medhasananda (2002: 441) discussed the constant inflow of settlers, growth in the population and negligence in the planning of the settlements which also triggered the filling up of endangered water tanks and ponds. He further argued that cleaning of water tanks regularly and maintenance activities were never carried out during the colonial regime.[40]

The Benares: A Gazetteer (1909, 239) mentioned the exercise of filling up water tanks as follows:

> From an early period of (the), British administration attention has been paid to the improvement of drainage in this area. Many tanks have been filled up and the process is still going on, while proper channels have been constructed to dispose of the surface water, thereby preventing the floods which were once of a serious nature in times of heavy rainfall, and producing a marked amelioration of the sanitary conditions of the town.

In 1867 the municipal board was established for dealing with municipal issues such as drainage, water supply and roads, under the Indian Municipal Act of 1850 (Medhasananda 2002: 441). Richard Temple (1881: 321), the former Governor of Bombay, discussed the importance of sanitation and appointments of sanitary commissioners at local administrative offices. Temple stressed that the supply of water was an obligatory task of every local administration. Continuing the urban sanitary reforms, a centralized waterwork system was built in 1892 and the townspeople of Varanasi started getting 'pure and filtered water'. Modern initiatives such as centralized water supply, (combined) drainage system, railways, hospitals, etc., were pompously demonstrated as 'monuments' of colonial authority.[41] The 1909 Gazetteer mentioned about 6,000 house connections in the municipal area (Nevill 1909: 263).

Edwin Greaves of the London Missionary Society of Benares discussed the history, geography and administrative set-up in his book titled *Kashi the City Illustrious or Benares* published in 1909. Greaves visited many temple tanks and dedicated one whole chapter describing the settings, condition of many water tanks and the role of the municipality. Greaves also provided information about the municipal administration regarding the waterworks and drainage systems to improve the sanitation of the city; to which he added that Europeans are unaware of the normal conditions of cleanliness in the 'East'. In the chapter titled, 'Temples, Tanks and other Places

Figure 3.6: Bhadaini water works at Benares (Varanasi). Photograph by Madho Prasad in 1905. The British Library Board, London

of Interest', Greaves stated about the poor quality of water and overall dilapidated condition of many tanks and ponds. Greaves mentioned Beniyā Tālāba (retitled as Victoria Park) and appreciated the improvement efforts by replacing the filthy wetland into a park.

In the same period, another author Charles Phillips Cape (1910: 131) wrote about the diseases and ponds in the city:

> Now there are several ways of preventing this sickness. One is to kill the mosquitoes whenever you see them. Another way and a better (one) is to destroy their larvae. This can be done by pouring paraffin oil on the ponds where they breed. If possible (,) all ponds and stagnant water near houses should be drained off.

The colonial rule changed the urban morphologies of the city and converted it into a modern dry city by eliminating the stagnant water pools. The 'Benares Guide Map' published in 1928 (see Figures 3.7 and 3.8) signifies the transformation of temple tanks and ponds in terms of forms and reduced expenses.

Figure 3.7: Benares Guide Map, 1928 (second edition). Calcutta Map Office, Calcutta

The detailed map reveals the elimination of some of the seasonal channels and water bodies. However, the map interestingly draws the tiny streams flowing down into these surviving water bodies and their catchment areas. The Guide Map of 1928 also depicted the numerous wells and detailed meandering courses of the Varana River and Asi rivulet.

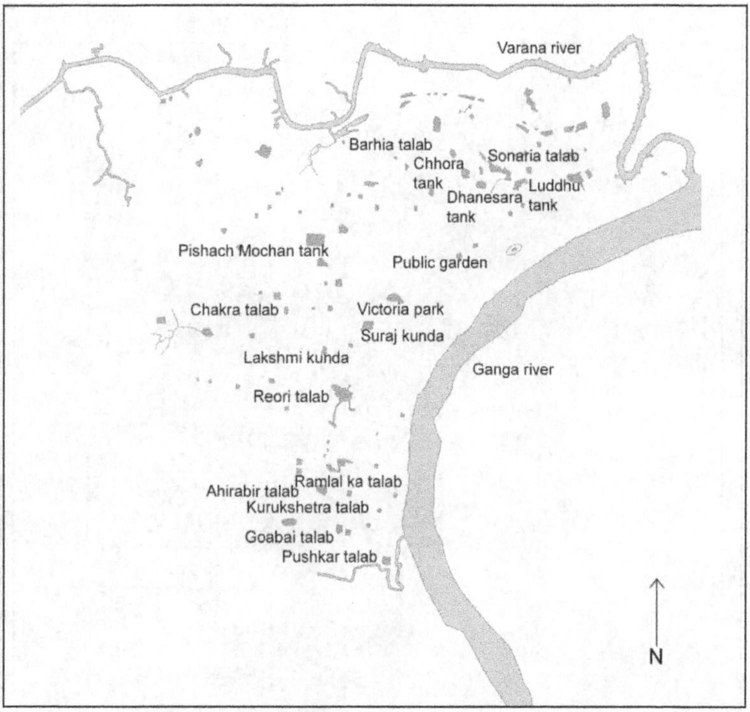

Figure 3.8: Tracing of the water bodies from the 'Benares Guide Map', 1928 (second edition). Calcutta Map Office, Calcutta.
Note: Map not to scale. The map made by the author

A careful comparison of James Prinsep's map of 1822 and the Benares Guide Map published after almost one hundred years reveals the authoritative implementation of urban sanitation policy by the British administration.

The 1928 guide map showed that names of many temple tanks and ponds were changed, and some names even indicated the colonial impression on it. For example, Mandākinī Tīrtha was mentioned as Mandākinī Tālāba in the 1822 map by James Prinsep and later the title was changed to Public Garden. Similarly, the earlier name of Veṇī Tīrtha was changed to Beniyā Tālāba and later appeared as Victoria Park on the 1928 guide map. Jyotīyā Tālāba was changed to Ahirabir Tālāba on the map of 1928 and this water tank later became extinct. Revā Tīrtha was mentioned as Revāḍī Tālāba and later became Revāḍī tank and is now extinct from the landscape of Varanasi. In 1892, the waterworks system was introduced, and the map of 1928 showed the existence of a 'Pumping Station' in the southern part of the city, near the River Ganga. Lord Curzon, during his service as Viceroy and Governor-General of India delivered a speech[42] in 1899, at Benares Municipality, in which he referred to the 'great scheme' of the waterworks which provided water and also immunity from the cholera epidemic. Curzon also showed grandiosity by referring to the extension granted by the government of the loan for the water-works scheme from 30 to 60 years. The detailed discussion of the James Prinsep Map of 1822 and the total number of extinct water bodies with the help of Georeferencing process is taken up in the following chapter.

CONTESTING THE COLONIAL URBAN 'REFORMS'

The strong arm of the colonial rule and the dominant urban reforms based on the Victorian policies related to sanitation, safety and modernity faced certain protests and also the solutions from the residents of Varanasi. A committee called '(Kāśī) Ganga Prasadhini Sabha' was formed with its main objective of the purification of the River Ganga. In a public meeting convened by the Maharaja of Benares in 1886 at the Town Hall, many notable personalities such as Raja of Durbhangha, Raja Shivaprasad, and many residents of Varanasi city discussed the drainage system of the city and the discharge of dirty water in the River Ganga (Muthiah 1911: 195, Wilhelm 2016). The committee approached the sanitary commissioner and suggested a new system of drainage to prevent the contamination of River Ganga.

Almost four decades later in 1926 'Kashi Tirtha Sudhar Trust' was established to improve the overall sanitation of the city, preventing the pollution of the River Ganga and restoring the *ghāṭas* on the western side of the river.[43] The president of Kashi Tirtha Sudhar Trust Motichand, in his preface (1931: Preface) articulated his concern and displeasure clearly over the indiscriminate discharge of sewage in the Ganga, which both hurt the Hindus minds and endangered public health. The trust also proposed to construct the parallel channel to the river stream and discharge the sewer further down in the river. The trust got appreciation from the colonial officers; the trust recognized the involvement and assistance of the people of Varanasi.

Raja Shivaprasad who was associated with the (Kāśī) Ganga Prasadhini Sabha also contributed to the initiative of starting a weekly journal called *Banaras Akhbar* from Varanasi in the mid-nineteenth century (Medhasananda 2010: 650). After this weekly journal, many other weekly journals were published from the city in Bengali, Hindi, Urdu and English. Swami Medhasananda (2010: 656) referred to the insufficient information that is available regarding the types of news and editorials that were published in Varanasi based journals. However, 'Selections from the Vernacular Newspapers Published in the Panjab, North-Western Provinces, Oudh, Central Provinces and Berar', 'Report of the Native Papers in Bengal', and 'Selections from the Native Newspapers Published in the United Provinces', provide valuable glimpses to the news published in the 'native' and 'vernacular' newspapers of the various provinces of India. The British government in India, collected, examined and translated these newspapers along with the information, such as the number of circulations, etc. and these bits of intelligence were treated as 'confidential' reports. These 'vernacular and native' newspapers discussed several issues from health, sanitation, pollution, social changes, traditions, customs, local taxation, trade and policing. A careful review of these newspapers reveals how the editors of these local and regional newspapers were voicing their opinions, offering suggestions and also submitting their protests against the colonial rule. The newspaper titled the 'English Hindosthán (Kalakankar)' discussed the poor condition of the 'sacred' temple tanks of Benares and also suggested a 'scientific' method to clean these water tanks.

The edition of 1 December, (1905) said:

> it is (an) open secret that the sacred temples and tanks the devout Hindus periodically visit are in most cases found unclean from a sanitary point of

view. (For) the bathing of millions of healthy and unhealthy persons, as well as the offerings of leaves and rice after being decomposed, makes the water of the narrow and small kunds (tanks) at Benares and other places nothing but stagnant. The devout pilgrims eagerly bathe in this water, sip and sprinkle it all over their body. This practice is, in our opinion, responsible for spreading diseases among the pilgrims who happen to visit those sacred tanks and dive in them; and therefore (,) we suggest that they may be repaired on scientific method with inlet and outlet so that they may continually receive fresh water and allow the old to escape.[44]

The newspapers published in Varanasi regularly complained about the unhygienic conditions of the city and also mentioned the poor condition of specific areas. These newspapers also discussed public health, lack of sanitary facilities, the spread of a pandemic and articulated their objections on several policies and actions of the imperial rule. These efforts were significant when the journal and newspapers often faced many challenges to continue their publishing endeavours (Medhasananda 2010: 656).

CONCLUSION

This chapter focused mainly on the transformation of various types of water reservoirs and discussed how the colonial reform policies had significantly altered the landscape, which was not clearly defined by dry and wet. The British administration started implementing urban reform policies by widening the roads and building the municipal gardens and drainage system by either reducing the size or drying up of temple tanks and ponds. The dominant reforms witnessed various forms of protests and contestation from the people. Many well thought out and earnest suggestions and recommendations were proposed by the residents of the city who enthusiastically and voluntarily participated in studying the changes and urban reforms. The reform and reclamation exercise affected mainly the natural inlets and outlets of many other water bodies and caused water to get stagnant and polluted. The colonial narrative of the dry city was considered safe and the task of filling up water bodies was set as a top priority to control the epidemics.

NOTES

1. Janine Wilhelm (2016) discusses how the well-established colonial ideas classified India as a 'tropical' environment and thus defended that a 'hot climate' will reduce the damaging elements. She also highlights 'India debt' along with other fiscal priorities of the British government that compromised with the effective implementation of the waterworks and sewage systems in India.
2. Edward Sachau, 1888, *Alberuni's India: An Account of the Religion, Philosophy, Literature, Geography, Chronology, Astronomy, Customs, Laws and Astrology of India About* AD *1030*, pp. 144-5.
3. Horton Ryley (1899: vii) described him as a 'pioneer Englishman' who brought practical information about India and the further east.
4. P.M. Kemp, (tr. and ed.), 1959, *Russian Travellers to India and Persia* (*1624-1798*), Kotov-Yefremov-Danibegov, p. 113.
5. In the 1803 Journals of the House of Commons (vol. 41) the Varanasi province is described as the seat of the Hindu religion. The report stated, 'the Territory of Benares is fruitful, and has been, not long since, an orderly, well cultivated, and improved Province', pp. 487-8.
6. Dean Mahomet also spelt as Din Mohammed, was serving in the East India Company. He travelled (1769-83) with an Anglo-Irish officer, Godfrey Evan Baker to various parts of India before arriving in colonial Ireland.
7. Sake Dean Mahomet, 1794, *The Travels of Dean Mahomet: A Native of Patna in Bengal, through Several Parts of India, while in the Service of the Honourable the East India Company*, p. 164.
8. Reverend Hobart Caunter (1794-1851) in his travelogue had an interesting description of one of the 'sacred tanks'. During his travels, he saw many alligators in a sacred tank. He was shocked to see that alligators were not hounded but fed by 'fanatic' Hindus Hobart Caunter, 1834, *The Oriental Annual, or Scenes in India*, p. 102.
9. F.S. Growse who was the magistrate and collector of Mathura wrote a book titled *Mathurá: A District Memoir* (1883). He wrote this reference book, especially for the district officers, described the numerous masonry tanks of Mathura, which were built with superior architectural skills.
10. Thomas Babington Macaulay wrote a lengthy essay on Warren Hastings (Governor of Bengal). In this essay, he give the colonial version of the events and the contribution of Warren Hastings that finally resulted in the complete annexation of Varanasi. See: Joseph Villiers Denney, (ed.), 1907, *Macaulay's Essay on Warren Hastings*, Boston: Allyn and Bacon.
11. Wilton Oldham, 1876, *North-Western Provinces: Historical and Statistical*

Memoir of the Ghazeepoor District, Part II: *History of Ghazeepoor and the Benares Province from* AD *1781 to* AD *1795*, p. 207.

12. *The Imperial Gazetteer of India, Provincial Series. United Provinces of Agra and Oudh*, 1908, vol. I, p. 124.
13. *Calendar of Persian Correspondence: Being Letters, Referring Mainly to Affairs in Bengal, Which Passed Between Some of the Company's Servants and Indian Rulers and Notables, 1792-3*, 1959, vol. X, Delhi: National Archives of India, p. 266.
14. With all the shortcomings the first census of Great Britain was undertaken in March 1801. The quantification exercise in Great Britain probably made an impact on the activities of James Prinsep.
15. *Supplement to the Indian Journal of Medical Research: Proceedings of the Third All-India Sanitary Conference*, 1914, Lucknow, p. 318.
16. David Kopf (1969) highlighted James Prinsep's contribution in Varanasi by mentioning street widening work and building the first modern sewage system of the city, p. 265.
17. H.R. Nevill 1909: 254, categorized the Matsyodarī tank as 'marsh' and gave credit to James Prinsep for converting it into the municipal garden.
18. As per the Kāśī Khaṇḍa the temple of Kāmeśvara was established by Sage Durvāsas, *SP* KKh. 33.120-2.
19. While investigating the urban landscape of Varanasi, Amita Sinha (2020: 259) defined the *ghāṭas* as the liminal place between the celestial River Ganga and sacred land.
20. The plan of constructing a tunnel for draining the Matsyodarī Tālāba was proposed in April 1825, and within two years in 1827, the masonry tunnel was completed.
21. E.W.C. Sandes (1935: 42) in his 'The Military Engineer in India' recounted an interesting account. He wrote that, when James Prinsep was working as Assay master in the Calcutta Mint, he opposed the plan of building a large masonry sewer at Hugli (now Hooghly) river to carry all the sewage of the city into Salt Lake. James Prinsep and his brother Captain Thomas Prinsep insisted that irregular flow of water would be dangerous to health.
22. For a discussion of the 'Nawabi drains and sewage disposal' see R.L. Singh, *Banaras: A Study in Urban Geography*, pp. 110-12.
23. United Kingdom (UK) Parliamentary Collections mentioned the 'Report on the Sanitary Condition of the Labouring Population of Great Britain' and appreciated the principal role and contribution of Edwin Chadwick. (https://www.parliament.uk/)
24. Feeding the fish and turtles is still common in Varanasi and devotees and local people regularly offer food to fish and turtles.

25. Dr Thomas Southwood Smith wrote an anonymous article for the *Westminster Review* (1825) and claimed that the 'certain condition of the air and atmosphere' instigated the widespread diseases.
26. James Hanley, in his article on 'Edwin Chadwick and the Poverty of Statistics' (2002) however argued that the statistical data of Edwin Chadwick's 1842 report was rejected by the statisticians and thus overlooked by the historians.
27. In India, during the colonial period, some of the attempts to investigate the causes of cholera and fever also marked the period from August to September as virulent. F.H. Fisher and J.P. Hewett, 1884, *Statistical, Descriptive, and Historical Account of the North-Western Provinces of India*, vol. XIV, Part I, p. 99.
28. 'Cholera caused havoc during the nineteenth century killing around 15 million population in India', David Arnold, 2000, *The New Cambridge History of India: Science, Technology and Medicine in Colonial India*, p. 81.
29. David Arnold (2000: 51) marked the 1820s decade as a significant shift where the initial constructive assessment of India was replaced by perceiving the barbaric and freakish environment of the tropical region. In the initial period, the assessment of the weather was also unsettled. The 'Report from the Select Committee on the Affairs of the East India Company' published in 1832 by the House of Commons (Britain) described the country's climate is mild and warm.
30. Tulchinsky, H. Theodore and A. Elena Varavikova, 2009, *The New Public Health: An Introduction for the 21st Century*, pp. 13-14. Arnold (1993: 194-5) referred to the discovery of cholera bacterium in India by Robert Koch, a German bacteriologist. However, David Douglas Cunningham and medical authorities stuck to 'anticontagionism' and set aside these results.
31. *Report of the Select Committee on the Affairs of the East India Company: With Minutes of Evidence in Six Parts and an Appendix and Index to Each*, 1832, House of Commons, Great Britain.
32. Ibid., p. 52.
33. During the time of Warren Hastings serving as first Governor-General (1774-85), the East India Company was enjoying the sovereignty of Bengal, Behar, Benares and Orissa. It also had a jāgirī of Madras and Bombay. (*The Contemporary Review*, vol. XXXII, April-July, 1878, London: Strahan and Company.)
34. The missionary work of the Church Missionary Society (CMS) was started in 1813 in the United Province. The CMS work in Varanasi was commenced in 1817 (Blunt 1912: 146-7).
35. Edwin Greaves joined the London Missionary Society and continued working in Varanasi for many years. Based on his 'personal enquiry and

observation' he wrote a book titled, *Kashi: The City Illustrious Or, Benares* (1909: 50). In this book, he has used the term 'East is East' to emphasize the lethargic attitude and delays of the natives in completing the work.

36. *Uttar Pradesh District Gazetteers*, Varanasi, 1965, pp. 70-1.
37. Report of the Commissioners Appointed to probe into the Sanitary State of the Army in India; with Abstract of Evidence, and of Reports received from Indian Military Stations, 1864.
38. Cohn (1962: 182-3) referred to F. Wilford, who was appointed a surveyor by the East India Company and surveyed Varanasi in the late eighteenth century. Wilford mentioned the filth and stink at some important water tanks and the dreadful condition of the drainage system.
39. *Calendar of Persian Correspondence: Being Letters, Referring Mainly to Affairs in Bengal, Which Passed Between Some of the Company's Servants and Indian Rulers and Notables, 1792-3*, 1959, pp. 819, 887, 1458.
40. Tristan Gooley (2016: 111) discussed the vulnerability of ponds and lakes. He explained that a lack of attention and maintenance would quickly turn the ponds and lakes into dry land. The negligence regarding a pond or a lake would let sediments gather and lead to turning in the wet mud, which would finally transform into dry land.
41. India in 1892, Indian Medical Gazette, 1898, p. 90.
42. Speeches by Lord Curzon of Kedleston Viceroy and Governor-General of India, vol. I, 1898-1900.
43. *Benares and its Ghats*, 1931, Kashi Tirtha Sudhar Trust, Benares.
44. Selections from the Native Newspapers Published in the United Provinces, No. 49 of 1905, p. 418.

CHAPTER 4

Debating Heritage Ecologies

This chapter draws on the previous chapter and the discussion on the reclamation agenda of colonial urban planning. This chapter maps and focuses on some of the temple tanks to study the transformation of waterscape that occurred mainly after the annexation of Varanasi by the British. The inquiry of these few selected temple tanks largely epitomized how the imperative urban imagination and its makeshift execution disrupted the catchment ecology by drying up the water bodies.

The opening part of this chapter focuses on some of the case studies of the selected temple tanks. These selected temple tanks exemplify how they were subtly described in the spatial texts and after that arranged and illustrated on the pictorial maps. The exercise of mapping has came to a full circle by locating the temple tank on the terrestrial landscape of Varanasi. These case studies also shed light on the trajectory of how once these active and functional temple tanks described as 'liminal places' dried up, encroached, turned into a cesspool and were finally eliminated. However, a few tanks as one sees, re-emerged with modified forms. It certainly restored the temple tanks in their modified and contemporary forms. However, this contemporary reclamation and beautification exercise did not necessarily restore the catchment areas, vegetation and open space. This restoration and beautification process in return demanded regular human involvement and structural intervention for the conservation of the water bodies. Therefore, it addressed and sustained the heritage yardstick but fell short on the various ecological thresholds.

In the next part of this chapter, the total number of titled temple tanks of Varanasi depicted on the map of 1822, prepared by James Prinsep, are enumerated. By applying the georeferencing method and physical mapping exercise conducted between February 2016 to November 2019 to understand the spatial transformation of the landscape by limitating the area mapped by James Prinsep on his map. In short, the study of the drying up and reclamation of water bodies and catchment ecology did not cover the present area of Varanasi city. The physical mapping details are also corroborated with the data gathered through direct observations, periodic surveys, interviews and discussions to understand the changes.

RECLAMATION AND REVIVAL OF WATER BODIES

The City Development Plan for Varanasi, 2041 (2015) mentioned the 29 important water bodies 'in' Varanasi (city). Except for three water bodies, the remaining water bodies were depicted on the 1822 map of James Prinsep. The development plan identified a total of 100 temple tanks and clarified that 12 temple tanks are extinct.[1] More than 10 per cent of the total slum areas of the city are located around these water bodies[2] and thus inimical to the slum dwellers and also pose a threat to water bodies. The deteriorating condition of the city's sewage system causes untreated sewage to flow into the surviving water bodies.[3]

RESTORATION OF MĀTṚ KUṆḌA

Chapter 97 of the Kāśī Khaṇḍa enumerated the sacred places of Varanasi. In this chapter, there is a reference of the Mātṛ Tīrtha and not specifically as a *kuṇḍa* (*SP* KKh. 97.245-6). Kāśī Khaṇḍa said that this site was located towards the north direction of Daśāśvamedhika. Sukul (1977: 94, 337) mentioned about the Mātṛ Pīṭha and a temple tank known as Mātṛ Kuṇḍa or Mātṛ tīrtha located to the western direction of Pitṛ Kuṇḍa.

James Prinsep, in his 1822 map, depicted the natural boundaries (not the rectangular shape with the leading stairs) for this water tank and titled it 'Matr Koond'. Prinsep located this water tank next to the 'Pitr Koond' on the west side. Prinsep's map also said that the environs of the Mātṛ Kuṇḍa sites were unenclosed. Edwin Greaves (1909: 90) located this water tank 'immediately to the west of Pitṛ Kuṇḍa and described its awful condition as a few steps were endured as the remanent of the temple tank.

When Sukul wrote his book in 1977, this tank was already extinct. Interestingly, Sukul's great grandfather Pandit Kailas Nath Sukul (1876 map) and Pandit Krishnachandra Sharma (1877 map) both categorically depicted Mātṛ Kuṇḍa (a rectangular form with stairs) to the north-west of Pitṛ Kuṇḍa. In the early twentieth century, when Edwin Greaves visited this temple tank, it was in poor condition. Gutschow claimed that this temple tank probably dried up in 1940 and after almost fifty years in 1996 an entrance gate was constructed to remember the extinct sacred site.[4]

Figure 4.1: Mātṛ Kuṇḍa gate.
Notes: Photo by Niels Gutschow in December, 2001, Benares; The Sacred Landscape of Varanasi, 2006

Figure 4.2: Mātṛ Kuṇḍa entrance gate
Photo by the author, May 2019

In 1990, a local committee called 'Mātṛ Kunda Samiti' was registered and restoration of a few temples was accomplished inside the complex of Mātṛ Kuṇḍa. The 'Google historical imagery' of 2001, displayed the dry open land without any marked boundaries and structures. Later in 2008 Varanasi Development Authority carried out the restoration work of the temple tank. The satellite imagery of 2008 and recently from 2020, reaffirmed the restoration exercise and the square water tank is visible. Currently, the water tank is surrounded by walls and compounds. The newly-built stairs lead to the water, which has many turtles and fish.

Figure 4.3: Mātṛ Kuṇḍa *Note*: Google historical imagery in March 2001

Figure 4.4: Mātṛ Kuṇḍa *Note*: Google historical imagery in December 2008

Figure 4.5: Mātṛ Kuṇḍa. *Note*: Google historical imagery in September 2020

Figure 4.6: Mātṛ Kuṇḍa. Photo by the author, August 2018

Figure 4.7: Temples near the Mātṛ Kuṇḍa. Photo by the author, August 2018

Figure 4.8: Mātṛ Kuṇḍa. Photo by the author, August 2018

The City Development Plan for Varanasi, 2041 (2015) stressed the mapping of water bodies and the removal of encroachments. The development plan also encompassed the need for improving the surviving water bodies to improve the ground water recharge and recommended creating pits for rainwater harvesting. However, unnecessary paving of open spaces and adding new structures inside and outside the water bodies went against the policy advice.

RESTORATION OF LĀṬ BHAIRAVA (KAPĀLMOCANA) KUṆḌA AND LAKṢMI KUṆḌA

Both Lāṭ Bhairava Kapālmocana tank and Lakṣmī Temple tank are well-known temple tanks and visited by both local persons and pilgrims travelling from other parts of India. The colonial authors discused the historical and archaeological dimensions of the Lāṭ Bhairava Kapālmocana tank in detail. The water tank is in the northern part of the city near to Varana River. James Prinsep depicted the square water tank located on one side of the small street and marked some vegetation around the tank. 'Banaras Guide Map' of 1928 and 1933 indicated the Lāṭ Bhairava Kapālmocana tank.[5]

Sherring (1868: 190) investigated the ancient pillar located near the water tank.

> To the north of the road leading from the Raj Ghat Port to the cantonments, at a distance of from three (-) quarters of a mile to a mile from the former place, is the Kapilmochan Tank. It is also called Bhairo ka Talao, or the tank of Bhairo. This is a strong and well-built structure, the stairs and foundations being of solid stone. On the high ground to the north of the tank stands a pillar, from seven to eight feet in height, and three in thickness, situated in the midst of a slightly elevated stone chabutra or platform. This is the Lat or pillar of Siva.

Sherring recounted the demolition of the temple by Aurangzeb and described the furious clashes that occurred during the early nineteenth century between Hindus and Muslims. He mentioned the decrease in the number of devotees and pilgrims frequenting this sacred temple site and water tank.

The tank fell into disrepair and was partly filled by debris. The stone steps were missing, and water was stagnant. In 2016 the reclamation and restoration work of the Lāṭ Bhairava Kapālmocana tank was initiated. The restoration work at Lāṭ Bhairava Kapālmocana tank and Lakṣmī Temple tank including the other two temple tanks were funded by the Corporate Social Responsibility (CSR) project. The restoration work was supported by one public sector enterprise of the Government of India and Varanasi Municipal Corporation. The works mentioned under the title 'Cleaning and Beautification', broadly included construction of barriers and entrance gate, the reparation of the walls and the approach roads, installation of soak pit and LED lights, purification of water, etc.

During fieldwork in August 2018, the tank was transformed and was surrounded by high walls. The cleaning and dredging activities were carried out to remove the silt, debris and garbage and clean the bed of the water tank. The four entry gates were

Figure 4.9: The hoarding mentions the cleaning and beautification work of the Lāṭ Bhairava Kapālmocana tank
Note: Photo by the author, August 2016

installed and the limestone steps were mounted by retaining the former stepping stair architecture (see Figure 4.5). A few water fountainheads are installed in the tank to circulate the water and for avoiding the process of stagnation.

Every morning and evening gates are open, and residents can walk on the surrounding expansive terrace of the tank. The renovation and beautification of the tank reclaimed the area and redefined the boundaries. Direct access is curtailed and possibly protects the tank from getting polluted by the garbage.

RETREATING TO WATER BODIES 'IN' THE CITY MODEL

Before the commencement of the 'Heritage City Development and Augmentation Yojana' and 'Smart Cities Mission', the restoration exercise was broadly confined to the activities such as constructing

August 2016

August 2016

February 2017

February 2017

August 2018

August 2018

November 2019

November 2019

Figure 4.10: Transformation of Lāṭ Bhairava Kapālmocana tank, Varanasi (from August 2016 to November 2019)
Note: All the photographs by the author

the pathway to the water tank site, or installing a railing or barrier along with the water pool, etc. The hoardings displaying the nature of renovation work carried in 2013-14 are still visible near some of the water tank sites. These hoardings announce that all ponds and water pools are 'properties' of the municipal corporation and warn that any kind of encroachment, that changes the existing forms is a punishable offence.

The ongoing restoration exercise of temple tanks and ponds focuses mostly on reclaiming the encroached water bodies, cleaning, removal of silt and debris, installing the walls, making the space for idol immersion, installing fountains, etc. Efforts such as repairing the architectural elements and an approach of making these temple tanks 'user-friendly' will answer the primary concerns. The restoration work is also arguably essential for the water bodies which are on the threshold of constant alterations and negligence. The construction work may protect from encroachment and relatively curb the easy access to the temple tanks and ponds. However, this effort might disrupt and undermine the organic features and functions of these heritage water structures. Another prominent feature of the restoration work is visible through the depiction of the landscape planning and graphical representation of the work. The landscape design plan of 'Matsyodarī Tālāba' restricts the earlier form of the water pool. The ongoing work reveals that the size of the water has body reduced relatively.

The review of restoration programmes and policies underlines the ecological features of these water bodies. However, the ongoing restoration and beautification work subscribes largely to the model of ecology 'in' the cities. It appears that the beautification agenda dominates the planning and execution somehow and thus does not engage with the interconnectivity with the macro urban ecosystem.

SPATIOTEMPORAL TRANSFORMATION

James Prinsep, depicted a total of sixty-one[6] titled temple tanks and ponds (see Table 4.1) on his map of 1822. In the last two centuries, thirty-three water bodies (out of the sixty-one titled water bodies) are now extinct, and on some of the sites, permanent constructions has taken place. If one compares the existing condition of the water bodies with the expanse and form of the water bodies depicted on the 1822 map what one finds is the profound geomorphic transformation of the city. The presence of water bodies declined radically along with the loss of catchment ecology, open space and connecting seasonal streams. After the reclamation process that started way back in the early nineteenth century, many water bodies were converted into streets, built-up areas, or part of the gardens. Interestingly, a few water bodies, such as Godauliyā Nālā (now, Godauliya road), and Misir Pokharā (now Mishra Pokhara) still appear as landmark locations and part of the postal addresses.

The contemporary urban city has a busy main road called Godauliya road and Dashashwamedh road where earlier there was a significant stream draining the excess water of inner water tanks and ponds such as Misir Pokharā, Hinguā Tālāba, Lakshmi Kuṇḍa, etc. into River Ganga (Singh 1955). This area is a low-lying area next to the Chauk area, which is an embankment.

Figure 4.11: The address of a building mentions the name Misir Pokharā (now Mishra Pokhara) as a local address
Note: Photo by the author, August 2016

TABLE 4.1: STATE OF THE TEMPLE TANK, PONDS IN VARANASI

Total Number of Titled Temple Tanks	Extinct	Restoration Activity	Relatively Good Conditoin	Natural-Open Boundaries	Polluted Dry Bed, Garbage, Encroach-ment
61	33	8	2	8	9

Note: Based on the fieldwork and physical mapping exercise from February 2016 to May 2019 at Varanasi. All the untitled water bodies are extinct.

The restoration and beautification work is taking place at the eight water tank sites. Among the existing water bodies, eight water tanks have natural or open boundaries and are therefore vulnerable in nature. Also, nine water bodies are highly polluted and facing problems such as garbage dumping, wastewater disposal and encroachment. Two temple tanks are close to the River Ganga and have a direct impact on the rise in the water level of River Ganga and deposition of silt. Therefore, these two temple tanks call for intense silt removal work after the excess water of River Ganga recedes. Also, since these two temple tanks are part of the Ghāṭa architecture, they receive huge attention and visits from the devotees and pilgrims.

CONCLUSION

This research has revealed that to the area mapped by James Prinsep in his 1822 map more than 54 per cent of the water bodies have been extinct in the last two hundred years. The elimination of a significant number of water bodies and their catchment areas is starkly visible during the events of floods. Flooding is a common phenomenon in these vast flood plains and is often described as an 'annual feature' (Joshi 1965: 137) when the surplus water enters into the low lying areas of the city and often advances in the other densely populated areas. The excess water remains in these areas despite the rainfall. The local and regional newspapers cover the flood and (re)draw attention to the encroachment and clogged

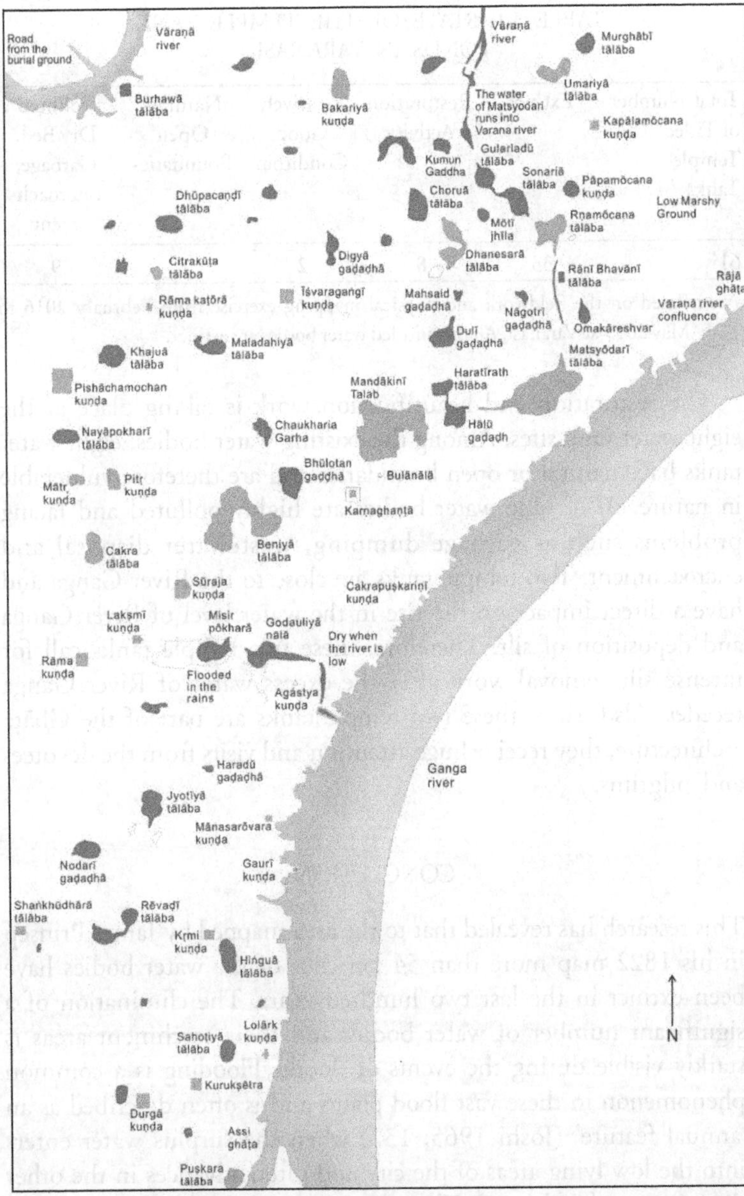

Figure 4.12: Tracing of James Prinsep Map of 1822 to highlight the existing water bodies and extinct water bodies based on the author's fieldwork

Note: Map not to scale and made by the author

drains. The Varanasi Master Plan 2031 recommended one-fourth area of the entire Varanasi development region be preserved as an open space and for 'urban forest'. However, today the majority of the surviving water bodies are surrounded by dense built-up areas which pose a threat to the surviving water bodies and their regular maintenance activities. The development plans and mission programmes evoke the 'of' understanding as a core principle in the rejuvenation of the water bodies. However, various gaps and limitations restrict the reclamation and revival of surviving water bodies to a limited extent.

NOTES

1. City Development Plan for Varanasi, 2041 (Final City Development Plan), 2015, Ministry of Urban Development, Government of India.
2. City Development Plan for Varanasi, 2041 (Final City Development Plan), 2015, Ministry of Urban Development, Government of India, p. 154.
3. There are total of 6 sewage treatment plants functioning and according to the Operational Guidelines for HRIDAY (2015: 138) around 130 MLD sewage is discarded untreated.
4. Niels Gutschow, 2006, *Benares: The Sacred Landscape of Varanasi*, p. 140. Gutschow in his book furnished the photograph (taken in 2001) of the entrance gate.
5. However, the guide map of 1928 and 1933 did not mention the title of the water tank.
6. Sukul (1977) has mentioned more than sixty water bodies on James Prinsep's map of 1822 but, does not reveal the precise number. The author has counted and added the 'Agastya Kuṇḍa' in the list-making the total number of sixty-one. The James Prinsep marked the place of Agastya Kuṇḍa and, therefore one considered it while preparing the list.

Conclusion

This book has offered the first detailed analysis of the early colonial draining of the water bodies followed by the evolving exercise of rejuvenation of extinct and surviving water bodies of Varanasi. This research has many uncharted areas which need to be explored and studied. Taking this into account, the following conclusion aims to connect the dots from the preceding chapters and discuss the findings and scope for further research. The fundamental assertion of this study, set forth in the introduction, is that the beginning of colonial arbitrary and experimental large-scale draining of temple tanks, ponds and streams formulated during the early East India Company regime, which was intensified during the British regime, caused the elimination of water bodies and loss of catchment ecology. This study has reassessed the hydrological transformation of the city in the realm of 'heritage ecologies'.

The findings of the first imperative process of modernizing Varanasi have demonstrated that colonial urban imagination and policies stemmed from Victorian sanitation policies, safety priorities and urge for modernity which made an enormous impact on the hydrological system of Varanasi. By evaluating the colonial dominance, which lasted for almost two centuries, this research has marked the early attempts of the draining exercise that triggered the pollution of River Ganga and eventually threatened the inland water bodies and streams.

The second process of transformation commenced soon after Independence of India. The Master Plans and city development initiatives were largely unsuccessful to reposition the urban imagination of the 'riverfront city' by mapping and exploring the surviving water bodies and threatened catchment areas. In such circumstances, the 'holistic development programme' missions marked the departure from their precursor development agendas. The two mission programmes (HRIDAY and SCM) announced comprehensive rehabilitation of the water bodies. The mission programmes dissociated themselves from the notion that aimed at the renovation of specific and independent water bodies in the city. This is described as 'heritage ecologies' which is an ongoing exercise and will continue to evolve and rehabilitate the surviving water bodies of the city.

This research has adopted the ecology 'in' and 'of' the city framework to analyse how an ancient city and its geography was disconnected from the water bodies and its catchment ecology and repositioned merely as a riverfront city. In this complex process of transformation, the majority of water bodies and seasonal streams were either eliminated or transformed to such an extent that they are now mere water containers without their catchment areas. The findings are, to some extent, compatible with the studies that delved into an exhaustive enquiry of river pollution and the elimination of water bodies and catchment ecology of the historical cities in India. However, the salient feature of this study is to assess mainly three eras with the 'in' and 'of' paradigm to locate how the hydrological landscape of Varanasi coped, adapted and adjusted with the external changes, policies and preferences that overrode other interpretations. The 'heritage ecologies' notion applied to retrospect the previous trajectory of the alterations and eliminations of water bodies and their traits in the ongoing restoration work.

The ecology 'in' and 'of' the city paradigm presented the notional spherical rotation of the water bodies and began with the water 'of' Varanasi. The opening of the book, reconstructs the environs of the city through the readings of descriptive texts. The broad consensus is extracted through these texts, and it reveals the omnipresence of water manifesting the 'of' notion. There is also an absence of an impulse to segregate the water and land to understand the sacred landscape. The water tanks and pools were named diversely, and the same temple tank bears differing names on different occasions revealing the kaleidoscopic nature of the water cycle. The notion of 'sacredness' which is innately unified with the abundant flowing water that consecrates the landscape was flagged. The composers of the Purāṇas, such as Kāśī Khaṇḍa were aware of the moving nature of the water in various forms. Probably, that is why they envisioned the 'Matsyodarī' as tīrtha, *tālāba* and a stream creating a seasonal, temporary island. The 'water sensitive' approach and treatment appreciated the rising water and forming of new fluid water bodies which will later bear the moist and wet but not the water. The introductory discussions are focused on the geography and locations of the temple tanks, ponds and streams—which are an integral part of the temple complexes. Before this study, the temple tanks and ponds were not investigated in the light of the spatial descriptive texts and their description. With the quantitative method of the Kāśī Khaṇḍa, the chapter has enumerated the list of 281 diverse water pools and ponds with their geographical locations. The descriptive texts also analysed with the pictorial mapping exercise of the late nineteenth century which used the Kāśī Khaṇḍa extensively to draw a panoramic view of the city. The chapter draws attention to more study of pictorial maps in the light of Purāṇas and especially the Bhuvanakośa sections.

The preparatory discussion on the spatial descriptive texts and pictorial maps is expanded further by bringing the first topographical map of James Prinsep published in 1822. This map is referred to as a base map in the entire study to understand the attributes from pre-colonial topography and to collate it with the transformation that happened mainly during the colonial regime. The Prinsep map of 1822 is notably significant as it represents the British cartographic notion of space in particular and the Western cartographic notion in general, which is exclusive and superior to other mediums and forms of representation. This map still carries the immense importance of its 'actual' representation of Varanasi. The two centuries old map deciphers the nature, environs, topography, edifices and other intangible dimensions by reinforcing the boundaries. This research has highlighted that James Prinsep made a remarkable contribution by mapping the seasonal streams, drainage patterns and revealing the slopes of the city. However, these detailed mapping efforts are inconsistent when he endeavoured to build the drainage system by overlooking the slope and embankment features and installing the 'combined system' of drainage. This research reemphasizes that the 'combined drainage' initiated the river pollution and contaminated the water bodies.

James Prinsep was active in Varanasi for ten years and these ten years were an essential landmark when the optics of modernizing Varanasi were founded. This research was limited on the primary sources of the sewage work and the other initiatives that were taken by the 'local improvement committee' and more in-depth research is required to shed light on the inclination of the members and make available other selections.

This research has a backdrop of recent unprecedented heavy downpour and massive flooding in many cities of India. The present research discussed the flood calamities to trace the mechanism thought and devised by the British officers in India. As the colonial forces advanced into the floodplains of River Ganga, the rising water was an unwelcome guest as flood brings cessation, fear and increases the trouble momentarily. Thus, the tool and measures of 'controlling the flood' become necessary. The dry solid ground overrode the wet, moist and unpredictable landscape and intensified the need for more ground to expand. The reclamation of swampy, marshy, non-functional and stagnant water pools, tanks were 'development' work that ensured hygiene and safeguarded from infectious diseases and an unsanitary environment. Although, this research is based on the limited investigation of the Victorian era of sanitary reforms to control and eradicate the infectious diseases owing to 'miasma', the findings have suggested their linkages and resonance in British India. The volatile events and the countermeasures taken in Europe were replicated in the missionary's accounts. The research also demonstrated how the missionary narrative shaped the colonial dispensation and reiterated the deteriorating conditions of temple tanks and ponds, which were directly responsible for generating the poisonous air, e.g., miasma.

Interestingly, some of the anecdotes from the accounts of the resident missionaries revealed the active and organic linkages between the temple and its water reservoirs. This research discerned how the elimination of a water reservoir seriously disturbed its connection with the temple and the frequenting devotees and pilgrims. The increasing number of deaths of British soldiers in India due to cholera, malaria, fever and other infectious diseases and the massive setback due to the 'War of 1857' alarmed the

Crown and British Parliament. These circumstances validated the large-scale implementation of sanitary reforms coupled with civic projects such as the widening of streets, municipal water supply, drainage projects, reclamation of swamps, marshy lands and ditches, etc.

The *British Medical Journal* published an 'appeal' of Dr Telafuss, with the title 'The Cholera'[1] in its 25 March 1893 edition, which had, according to the journal, studied the treatment of cholera. The journal also mentioned that his conclusions are corresponded with the editor of the journal. Dr Telafuss had raised three main issues. First, he categorically announced the delta of River Ganga and its larger territory as the 'home' of cholera. The heavy rains, tropical weather, wet and moist terrain increased the probability of pandemics. Second, he advised the British government to plant eucalyptus trees across the country so that the trees would not only absorb all the wetness and humidity but most importantly extirpate the miasmatic substance. Third and most importantly, he advised the British to build canals and drainage of the entire delta of River Ganga and Brahmaputra. Even though some of the concerns were restated and drainage exercise costing substantial financial burden on the British government were carried out, this appeal of Dr Telafuss effectively optimized the notions and theories on which the sanitary policies, sewage systems, drainage and reclamation strategies were crafted and implemented. The review of some of the colonial urban reforms that were rolled out in Varanasi pointed out the relevance of 'of' understanding. The colonial stakeholders worked primarily on the analogy model and imported the theories from home and implanted those combined with an attitude of superiority and a great amount of cautiousness. These urban reforms were also calibrated with the strategical importance of the places and picking the most cost-effective solutions. The 'Rivers Pollution Prevention Act of 1876' which was passed to maintain and restore the 'wholesomeness of rivers and other inland or coastal waters of England and Wales'[2] was not consider while introducing the drainage works at the 'sacred city of Varanasi'.

The transformation of the urban morphologies of Varanasi is highlighted by juxtaposing the James Prinsep map of 1822 and the 'Benares Guide Map' of 1928. The comparison of these two topographical maps reveals that during this period of hundred years, almost all of the water bodies and seasonal streams were either extinct or significantly reduced in size. In the closing segment of the book, the number of surviving water bodies confined to the James Prinsep map of 1822 is listed.

Although the study is based on a confined territory of the expanding city, the discussion and quantitative analysis suggest that the water 'of' Varanasi transformed from its ubiquitous free-flowing nature into the fixed boundaries both on the land and on paper. After 1947, the approach of the preservation and renewal of surviving water bodies was limited to the selected water tanks and ponds. The urban renewal mission discussed the 'development of heritage areas' and 'preservation of water bodies'[3] and remained in the 'in' paradigm overshadowed by the previous approaches. In 2014, Varanasi was clearly at crossroads, when it was selected for two national flagship missions: HRIDAY and SCM. Both these missions are vocal for the comprehensive rejuvenation of the city by bringing the heritage in the broader framework of planning tools and processes of urban local bodies. The mission programmes articulated the missing and neglected components in the previous attempts. They also highlighted the tapping of human actors and their aspirations, including tangible elements such as knowledge, skills and capabilities. The mission statements discussed the natural heritage in the context of its biodiversity and ecology. It brought these two concepts and enabling to craft the concept of heritage ecologies to discuss the water 'of' Varanasi. However, the periodic assessment and fieldwork studies of the ongoing process of heritage rejuvenation disclosed a different reality on the ground. Even though the process of reclaiming and rejuvenating the surviving water bodies was in progress, the exercise was confined mostly to the 'in' model.

It is intriguing to see how the work in progress evolves and unfolds in forthcoming times. It will be delusory to imagine and strive for the landscape as it was mentioned in the descriptive text of Purāṇas and other literature. The city has travelled a long distance from its abundant free-flowing waterscape to its modern creation of a riverfront city, and many water bodies and streams are not only hard to locate but extremely difficult to restore. This research, with its inquiry into the water bodies of the city, calls for more engaging and comprehensive research.

NOTES

1. Ernest Hart (ed.), 1893, *The British Medical Journal*, vol. 1, London: The British Medical Association, pp. 662-3.
2. The Rivers (Prevention of Pollution) Act, 1951, in its preface outlines the provisions of Rivers Pollution Prevention Act of 1876.
3. Guidelines for Jawaharlal Nehru National Urban Renewal Mission (JNNURM), December 2005, Ministry of Urban Development.

APPENDICES

APPENDIX I

Sacred Temple Tanks and Water Pools Stated in the Second Chapter of Varanasi Māhātmya of *Tīrtha Vivecana Kāṇḍa*

1. Amaraka Hrada
2. Aṅgāraka Kuṇḍa (KKh. 97.67)
3. Asi Nadī (KKh. 7.66)
4. Bali Kuṇḍa
5. Bhadradroha Hrada
6. Bhadrakālī Hrada (KKh. 97.156)
7. Ganga Varana Saṅgama (KKh. 33.100)
8. Ghaṇṭākarṇa Hrada (KKh. 53.37)
9. Kapālmocana Tīrtha (KKh. 31.131)
10. Kapilā Hrada (KKh. 62.47-53)
11. Kōṭi Tīrtha (KKh. 97.63)
12. Lakṣmī Kuṇḍa (Śrī Kuṇḍa) (KKh. 64.18)
13. Lolārka (KKh. 46.49-59)
14. Mandākinī (KKh. 64.10)
15. Mārkaṇḍeya Hrada (KKh. 83.107)
16. Matsyodarī (KKh. 33.120)
17. Pañcacūḍa Hrada (KKh. 97.146)
18. Vaitaraṇī (KKh. 64.13)
19. Varana Nadī (KKh.)
20. Varanasi Jānhavī Saṅgama (KKh.)
21. Vāsuki Tīrtha (KKh. 64.20)
22. Vinayaka Kuṇḍa (KKh.)
23. Vyāsa Kuṇḍa (KKh. 97.144)

Notes: The list of *tīrthas* and *kuṇḍas* from *Tīrtha Vivēcana Kāṇḍa Kṛtyakalpataru* of Bhaṭṭa Lakṣmīdhara is referred. The titles of rivers, streams, temple tanks and ponds are arranged alphabetically. The above titles mentioned by Lakṣmīdhara are verified with the list of Kāśī Khaṇḍa.

APPENDIX II

List of Temple Tanks and Water Pools (Excluding Wells) from Kāśī Khaṇḍa of *Skanda Purāṇa* Arranged in Alphabetical Order

No.	Title	Chaptes and verses
1.	Ādigadādhara Tīrtha	58.51
2.	Ādikeśava Tīrtha	83.127
3.	Ādityakeśava Tīrtha	84.17
4.	Agasti Kuṇḍa	61.179
5.	Agasti Tırtha	61.177
6.	Agastīśvara Kuṇḍa	25.48-9
7.	Agastya Tīrtha	61.80, 83.106
8.	Āgneya Kuṇḍa	97.120
9.	Agni Tīrtha	84.73
10.	Agnīdhra Kuṇḍa	100.63
11.	Aikṣvākava Tīrtha	84.70
12.	Airāvata Kuṇḍa	64.16, 97.134
13.	Aṁbarīṣa tīrtha	58.49, 61.12, 84.16
14.	Amṛtasrāva Gaṅgā (no category)	77.51
15.	Anāraka Kuṇḍa	97.60
16.	Aṅgāra Tīrtha	84.74
17.	Aṅgāraka Tīrtha	97.67
18.	Arka kuṇḍa / Barkari Kuṇḍa*	47.1, 47.54, 47.56
19.	Ārundhateyaka Tīrtha	83.107
20.	Arundhatī Tīrtha	61.168
21.	Aśōka Tīrtha	33.180, 83.92, 97.148
22.	Avadhūteśa Tīrtha	97.180
23.	Avimukta Tīrtha / Dharmanada	84.103, 59.97
24.	Avimukteśvara Tīrtha	61.117
25.	Ayogandha Kuṇḍa	69.22
26.	Bāṇa Tīrtha	84.35
27.	Bandī Kuṇḍa	97.136, 83.84, 83.88
28.	Bhadra Hrada	97.46

No.	Title	Chaptes and verses
29.	Bhadrakālī Hrada	97.156
30.	Bhadrakarṇa Hrada	69.104
31.	Bhadra Vāpī	70.44
32.	Bhagīratha Tīrtha	83.109, 83.127, 84.105
33.	Bhāgīrathī Tīrtha	61.157-9, 83.79-81
34.	Bhairava Tīrtha	84.46
35.	Bhārgava Tīrtha	58.52, 61.14, 84.19
36.	Bhava Tīrtha	61.202
37.	Bhavānī Tīrtha	61.123-39, 83.94 84.104, 100.71
38.	Bhīma Kuṇḍa	70.73
39.	Bhūpālaśrī Tīrtha	58.223
40.	Bhuvaneśvara (pit)	97.72
41.	Bindu Tīrtha	59.136, 59.140-1, 60.72-3; 115; 120; 124, 84.63, 100.71
42.	Brahma Nālā	61.150; 155-7, 69.131, 83.97, 84.105
43.	Cakra Hrada	97.51
44.	Cakra Tīrtha	58.36, 84.9
45.	Cakrapuṣkariṇī	94.31, 94.34, 100.39, 26.53,
46.	Cakrapuṣkariṇī Tīrtha	26.64-5, 61.83, 79.78, 84.86, 91.2, 96.11
47.	Candra Tīrtha	84.76
48.	Candroda Lake	14.50-1, 14.55-7
49.	Catuhsagara Vāpikā	66.17, 66.18,
50.	Catuḥsrotasvinī Tīrtha	76.149
51.	Caturnada Tīrtha	76.131, 79.98
52.	Caura tīrtha	83.67
53.	Citrarathesvara Tīrtha	83.100
54.	Dadhīcīśvara Tīrtha	97.11
55.	Dadhikalpa Hrada	97.184
56.	Dandakhata Tīrtha	64.5-9, 65.45, 97.170, 100.59
57.	Dārukeśvara Tīrtha	70.80
58.	Daśāśvamedha Tīrtha	52.68, 52.83, 52.84-87, 52.90, 52. 94, 52.100, 61.29, 83.83, 84.108

No.	Title	Chaptes and verses
59.	Daśāśvamedhika Tīrtha (original name was Rudrasaras Tīrtha; 52.69)	33.178, 97.244
60.	Dattātreya Tīrtha	84.18
61.	Dattātreyeśvara Tīrtha	58.51
62.	Dharma Tīrtha	81.1-2-22-33, 83.101
63.	Dharma Jalāśaya	81.28
64.	Dharmanada Hrada / Tīrtha	59.104, 59.138, 59.136, 59.142
65.	Dhruva kuṇḍa / Tīrtha	21.128, 64.19
66.	Ḍhuṇḍhi Tīrtha	61.122, 84.103
67.	Dhūtapāpā Tīrtha	59.139, 60.8-10, 60.137-8
68.	Dilīpa Tīrtha	83.63
69.	Dundubhinir Hrada	65.46-65
70.	Durgatisamhar Tīrtha	64.22
71.	Durvāsas Tīrtha	64.11
72.	Gadā Tīrtha	58.37, 84.10
73.	Gadādhara Tīrtha	62.63
74.	Gaja Tīrtha	83.61
75.	Gandharva Kuṇḍa	66.21
76.	Gandharvāpsaras Tīrtha	64.16
77.	Gaṅgā Keśava Tīrtha	61.180, 83.106
78.	Gapīgovinda Tīrtha	84.24
79.	Garua Tīrtha	83.96
80.	Gārutmata Tīrtha	84.13
81.	Gaurī Kuṇḍa / Tīrtha	77.47, 77.50
82.	Gautama Kuṇḍa	64.21
83.	Gautamī Tīrtha	83.105
84.	Gayā Tīrtha	69.36
85.	Ghaṇṭākarṇa Hrada	53.37, 53.40-1, 95.68, 95.71, 95.74, 97.144, 97.145
86.	Ghṛtakulyā (Kṛtakṛtyā Tīrtha)	62.73
87.	Gokarṇeśa Lake	18.15
88.	Gopīgovinda Tīrtha	58.57, 61.19
89.	Gopratāreśvara Tīrtha	84.36
90.	Gopreksa	100.68
91.	Govyāghreśvara Tīrtha	83.71

No.	Title	Chaptes and verses
92.	Haṃsa Tīrtha	64.11, 68.57-65, 77.47, 83.68
93.	Harapāpa hrada / Tīrtha	77.9-10-11-17, 77.43-44-48, 83.81
94.	Hariścandra Tīrtha	61.74-7, 84.80
95.	Hayagrīva Tīrtha	58.61, 61.23, 83.61
96.	Hayagrīveśvara Tīrtha	70.88
97.	Hiraṇyagarbha Tīrtha	84.37
98.	Hiraṇyagarbheśa Tīrtha	100.54
99.	Indra Kuṇḍa	97.116
100.	Indradyumna Tīrtha	84.68
101.	Īśāna Tirtha	33.42, 61.139, 84.104
102.	Jānakī Kuṇḍa	64.20
103.	Jarāsandheśa Tīrtha	83.103
104.	Jñāna Tīrtha	33.37, 33.39, 61.139, 84.104
105.	Jñānahrda Tīrtha	84.57
106.	Jñānoda Tīrtha	32.161, 33.1-33-38-44-46-49-76, 57.35
107.	Jyecmha Tīrtha	63.19
108.	Jyeṣṭha Vāpī	63.12-15, 66.5, 100.68
109.	Kālagaṅgā Tīrtha	84.67
110.	Kālatamodaka (no category)	24.76
111.	Kāleśvara Tīrtha	83.91
112.	Kali Tīrtha	84.75
113.	Kāloda Well	97.125,
114.	Kālodaka Well	24.74
115.	Kāma Kuṇḍa / Tīrtha	85.78, 97.101
116.	Kāmada Tīrtha	61.36
117.	Kāma Juṇḍa	85.73
118.	Kambalāśvatara Tīrtha	61.80
119.	Kambaleśvara Tīrtha	84.82
120.	Kapālmocana Tīrtha	31.131, 31.132, 31.138, 33.116, 64.12, 69.15, 69.112, 97.66
121.	Kapilā hrada / Tīrtha	62.47-53, 97.19,
122.	Kāpiladhāra Tīrtha	62.57-64
123.	Kāpileya Hrada	100.56
124.	Karkoma Vapī	66.23, 97.117
125.	Karṇāditya Tırtha	84.45

No.	Title	Chaptes and verses
126.	Kedāra (no category)	77.62
127.	Kedāra Kuṇḍa	77.54-56-58-59
128.	Kharvanṛsiṁha Tīrtha	84.48
129.	Khurakartari Tīrtha	61.161
130.	Khurakartari Tırtha	83.108
131.	Kokāvarāha Tırtha	83.62
132.	Komi Tīrtha	97.63
133.	Kṛtakṛtyā Tīrtha	62.62
134.	Kṣīrābdhi Tırtha	58.31, 84.8
135.	Kṣīranīradhi Tīrtha	62.62
136.	Kṣīroda (no category)	84.8
137.	Kṣonī Tīrtha	83.90
138.	Lakṣmī Tīrtha	64.18
139.	Lakṣmīnṛsiṁha Tīrtha	58.58, 84.25
140.	Lalitā Tīrtha	70.18-21, 83.103, 100.70
141.	Lolārka Kuṇḍa	46.49-53-59, 53-78, 61-25, 69-8, 96-201, 100-118
142.	Madhusravā Puṣkariṇī	62.62, 62.74
143.	Madhusravā Gaṅgā	77.47
144.	Madhusravas (no category)	62.81
145.	Mahā Kuṇḍa	97.96-119-31
146.	Mahākāla Kuṇḍa	97.136
147.	Mahālakṣmī Tīrtha	58.39, 84.12
148.	Mahiṣāsura Tīrtha	84.33
149.	Mahodadhi Tırtha	83.66
150.	Maitrāvaruṇa Tīrtha	84.72
151.	Makha Tīrtha	84.62
152.	Mānasa Saras (no category)	64.19 (not clear about the type of water body)
153.	Mānasa Tīrtha	77.52
154.	Mandākinī Tīrtha	64.10, 67-61, 67.170-194, 97.152-155, 100.54,
155.	Māndhātṛ Tīrtha	83.72
156.	Maṅgala Tīrtha	84.59
157.	Maṇi Kuṇḍa	68.82
158.	Maṇikarṇi	21.113, 61.106, 74.46, 100.55
159.	Maṇikarṇikā Tirtha	2.61, 3.6-8, 7.79-80, 10.79, 26.62-63-114-116-117-119,

No.	Title	Chaptes and verses
		30.3-80, 33.102-4-109-111-113, 34.5-8-27-28-31-32-33 34-35-105, 44.43, 57.35, 61.45-49-73-75-85-102-4, 75.54, 84.85-110, 94.21, 96.29, 99.50-51-54, 100.13-78-105
160.	Marīci Kuṇḍa	18.16
161.	Marīcīśa Kuṇḍa	97.116
162.	Mārkaṇḍa Tīrtha	61.164
163.	Mārkaṇḍeya Hrada	83.107, 97.104-7
164.	Marutta Tīrtha	84.71
165.	Mātṛ Tīrtha	97.245
166.	Matsyodarī Tirtha	33.120, 64.12, 69.21-136-37-39, 73. 155-58-60-162-65, 74.105-7, 97.79-82-110-12, 100.44
167.	Menakā (holy) Pit	64.15
168.	Mokṣa Tırtha	33.45
169.	Mṛkaṇḍa Tīrtha	84.49
170.	Mucukunda Tīrtha	83.73
171.	Mukti tīrtha	61.115-17, 84.103
172.	Mukuṭa Kuṇḍa	97.259
173.	Nābhi Tīrtha	61.153, 84.105
174.	Nāgeśvara Tīrtha	84.44
175.	Nandī Tīrtha	61.44, 84.104
176.	Nārada Tīrtha	58.45, 61.10, 84.14
177.	Naranārāyaṇa Tīrtha	58.54, 61.16, 84.21
178.	Narmadā (streams) / Tīrtha	61.172, 75.6, 83.107
179.	Nīlagrīva Tīrtha	84.28
180.	Nṛsiṃha Tīrtha	83.99
181.	Padma Tīrtha	58.38, 84.11
182.	Pādodaka Tīrtha	41.175, 51.75-76-77-81, 60.4-5, 61.4-7, 75.75, 84.3-5-7-109
183.	Paiśaṅgila (Pilipilā Tīrtha)	50.150
184.	Paitāmaha Tīrtha	61.149, 62.63, 84.104
185.	Pañcabrahmātman Tīrtha	33.151

No.	Title	Chaptes and verses
186.	Pañcacūḍa Hrada	97.146, 100.57
187.	Pañcanada hrada / Tīrtha	58.202, 59.5-6-11-13-17-126-27-43, 60.5-7-24-48-62-68-70-136-39, 61.1-43-246-49, 66.126, 67.168, 75.54, 79.102, 84.50-109, 97.183, 100.22
188.	Pañcatīrthī (no category)	84.114-107
189.	Paraśurāma Tīrtha	83.75
190.	Parvata Tīrtha	61.78, 84.81
191.	Pāśupata Tīrtha	61.106-11, 84.102
192.	Phalgu Tīrtha	33.35, 69.36, 81.31
193.	Pilipilā Tīrtha	55.26, 57.96, 69.167, 73.4, 75.12-19-20-47-67-72-74, 84.42, 92.16
194.	Piśācamocana Tīrtha	54.74-79-83-84, 69.67
195.	Piśaṅgila Tīrtha	57.87, 84.39
196.	Pitāmahasrotikā Stream	97.205-6
197.	Pitṛ kuṇḍa / Tīrtha	62.63, 64.19, 97.235
198.	Pitṛkūpa	97.92-123
199.	Prabhāsa Tīrtha	83.95
200.	Prahlāda Tīrtha	58.48, 61.11, 84.15
201.	Praṇava Tīrtha	84.38
202.	Prasannoda (holy) Pit	97.31
203.	Prayāga Tīrtha	33.179, 61.29, 69.39, 70.47, 83.89
204.	Pṛthu Tīrtha	83.74
205.	Pṛthūdaka (holy) Pit	64.14
206.	Puṣkara Tīrtha	69.20
207.	Ṛṇamocana Tīrtha	33.117, 64.13, 69.16, 97.66
208.	Rudra Kuṇḍa	97.91, 97.95
209.	Rudra Sarovara	52.88
210.	Rudrakoṭi Tīrtha	69.48
211.	Rudrasaras Tırtha	52.69
212.	Rudravāsa Tīrtha	61.111, 84.102, 97.89
213.	Sagara Tīrtha	83.64
214.	Śailādi Tīrtha	61.142
215.	Sāṃba Kuṇḍa	48.47

No.	Title	Chaptes and verses
216.	Sangamesvara Tīrtha	64.23
217.	Śaṅkha Mādhava Tīrtha	58.60
218.	Śaṅkha Tīrtha	58.35, 84.8
219.	Sankhacuda Kuṇḍa (Tank)	67.186
220.	Śaṅkhamādhava Tīrtha	61.22, 84.27
221.	Sāṅkhya Tīrtha	84.30
222.	Sannihati Puṣkariṇī	69.8
223.	Saptābdhi Tīrtha	83.66
224.	Saptasāgara Tīrtha	83.65
225.	Sārasvata Tīrtha	84.83
226.	Śaunaka Hrada	97.157
227.	Śaunakeśvara Kuṇḍa	100.59
228.	Śeṣa Tīrtha	58.59, 61.21, 84.26
229.	Siddha hrada / Kuṇḍa	69.124, 97.254
230.	Siddha Vāpī	97.165
231.	Siddhyaṣṭaka (Holy Pit)	67.173
232.	Śiva Tīrtha	33.31, 33.45-47
233.	Śivagayā Tīrtha	62.64
234.	Skanda Tīrtha	61.120, 84.103
235.	Śrīkaṇṭha Kuṇḍa	97.108
236.	Śrīkuṇḍa Tīrtha	70.63-64
237.	Śubha Kuṇḍa	97.120
238.	Sudhākhani Tīrtha	62.64
239.	Śukra Tīrtha	83.93
240.	Supratīka Lake	66.29
241.	Svalīna Tīrtha	84.31, 100.53
242.	Śvetadvīpa Tīrtha	61.8
243.	Takṣaka Kuṇḍa	66.12
244	Tāmra Kuṇḍa	97.58
245.	Tāmra-Varāha Tīrtha	84.66
246.	Tāra Tīrtha	73.175
247.	Tāraka Tīrtha	33.45, 61.118, 84.103
248.	Tārkcya Tīrtha	58.44, 61.9
249.	Tribhuvanakeśava Tīrtha	83.70
250.	Trisandhyā Tīrtha	61.174
251.	Trisandhyākhya Tīrtha	83.106
252.	Uddālaka Tīrtha	70.78, 84.29
253.	Ugra Kuṇḍa	97.114

No.	Title	Chaptes and verses
254.	Umā Tīrtha	84.84
255.	Urvaśī Kuṇḍa	64.15
256.	Vaitaraṇi	64.13(?), 97.94
257.	Vāmana Tīrtha	58.53, 61.15, 84.20
258.	Varāha Tīrtha	61.203
259.	Vasiṣṭha Tīrtha	61.166-172, 83.107
260.	Vāsuki hrada / Kuṇḍa	64.20, 66.8-9
261.	Vāyukuṇḍa	13.29
262.	Vetāla Kuṇḍa	68.74
263.	Vidāraṇa Narasiṁha Tīrtha	58.56, 84.23
264.	Vidhi Tīrtha	83.98
265.	Vighneśa Tīrtha	84.78
266.	Vimaloda Lake / Tīrtha	54.2-3-6-55, 97.73
267.	Vīra Tīrtha	83.111, 84.77
268.	Vīrarāmeśvara Tīrtha	84.69
269.	Viśāla Tīrtha	70.4-5, 83.102
270.	Viśālagaṅgā	100.70
271.	Viṣṇu Pādodaka	58.18-25, 84.4
272.	Viṣṇu Tīrtha	61.144, 84.104
273.	Viśva Tīrtha	61.113, 84.102
274.	Vṛddhārka Tīrtha	83.98
275.	Vṛṣabhadhvaja Tīrtha	62.60-63-68-78-82
276.	Vṛṣeśa Tīrtha	64.17
277.	Vyāsa Tīrtha	97.144
278.	Yajñavarāha Tīrtha	58.55, 61.17, 84.22
279.	Yakṣiṇī Kuṇḍa	64.17
280.	Yama Tīrtha	51.107-8-10-11-12-15, 57.66-112, 69.108
281.	Yoginī Tīrtha	61.176, 83.106

Source: Ganesh Vasudeo Tagare 1996, 1997, *Kāśī Khaṇḍa: The Skanda-Purāṇa.*

Notes: The English translation of the *Kāśī Khaṇḍa* has been the primary source for the study and for making the catalogues. However, the Sanskrit-Hindi translation text is additionally referred to verify the titles and other information.

APPENDIX III

List of Surviving and Extinct Temple Tanks and Ponds of Varanasi City Confined to the Area of James Prinsep's Map of 1822

No.	Names mentioned in the James Prinsep Map of 1822	Names from spatial texts and other scriptures	Present name and notes	Photograph
1.	Agastya Kuṇḍa (marking only below dotted marking line of Godauliyā Nālā)	Agastya Kuṇḍa		Extinct
2.	Bakariyā Kuṇḍa	Barkarī Kuṇḍa, Uttarārk Kuṇḍa, Arka Kuṇḍa	Barkarī kuṇḍa, Bakariyā Kuṇḍa Condition: Dry and polluted, encroachment	 August 2018
3.	Beniyā Tālāba	Veṇī Tīrtha	Beniyā Bāga Tālāba, Beniyā Park, Victoria Park Condition: Dry and polluted	 May 2019
4.	Bhūlotan Gaḍḍhā	Bhuleṭan Pōkharā	(Bhuleṭan Temple)	Extinct
5.	Bulānālā (marking only)	Vilok Tīrtha		Extinct
6.	Burhawā Talāba	Aśōka Tırtha		Extinct

No.	Names mentioned in the James Prinsep Map of 1822	Names from spatial texts and other scriptures	Present name and notes	Photograph
7.	Cakrapuṣkariṇ Kuṇḍa	Cakrapuṣkariṇī, Maṇikarṇikā	Maṇikarṇikā, Cakrapuṣkariṇī Tīrtha	 May 2019
8.	Cakra Tālāba	Cakra Talāba	Condition: Dry bed, polluted, encroachment	 August 2018
9.	Chaukuṅriyā Gaḍḍhā			Extinct
10.	Choruā Tālāba	Marīcḥ, Mārīcheshvar Kuṇḍa		Extinct
11.	Citrakūṭa Tālāba	Citrakūma Tālāba, Vināyak Kuṇḍa	Citrakūṭa Tālāba	 August 2018
12.	Dhanesarā Talāba	Dhanadeśvar Kuṇḍa	Dhanesarā Pōkharā Condition: Pollution, garbage, encroachment	 August 2018
13.	Dhūpacaṇḍī Tālāba	Dhūpacaṇḍī		Extinct
14.	Digyā Gaḍḍhā			Extinct
15	Dulī Gaḍḍhā	Mahākāleśvar Kuṇḍa		Extinct
16.	Durgā Kuṇḍa	Durgā Kuṇḍa	Durgā Kuṇḍa Condition: Restoration work	 August 2018

No.	Names mentioned in the James Prinsep Map of 1822	Names from spatial texts and other scriptures	Present name and notes	Photograph
17.	Gaurī Kuṇḍa	Gaurī Kuṇḍa, Haraṅpāpa Tīrtha, Kedāra Kuṇḍa	Gaurī Kuṇḍa, Condition: Contaminated	 February 2016
18.	Goābāī kuṇḍa (Untitled)	Goābāī Kuṇḍa, Mukuṭa Kuṇḍa	Guanbāi Gaḍḍhā, Goābāī Kuṇḍa	  August 2018
19.	Godauliyā Nālā	Gautama Kuṇḍa, Godāvarī Tīrtha	Godauliyā Street	Extinct
20.	Gularladū Tālāba		Gola Gaḍḍhā	Extinct
21.	Hālū Gaḍḍhā			Extinct
22.	Haradū Gaḍḍhā			Extinct
23.	Haratīrath Tālāba	Haṃsa Tīrtha	(Telephone Tower)	 August 2016
24.	Hiṅguā Tālāba			Extinct
25.	Īśvaragaṅgī Kuṇḍa	Agnīśvar Kuṇḍa	Īsvaragaṅgī Tālāba Condition: Restoration work, polluted, garbage	 August 2018
26.	Jyotiyā Tālāba		Ahirābīr Tank	Extinct

No.	Names mentioned in the James Prinsep Map of 1822	Names from spatial texts and other scriptures	Present name and notes	Photograph
27.	(*The LATH*) Kapālmocana Kuṇḍa	Lāṭ Bhairav, Daṇḍapāṇi Bhairav Tīrtha, Kapal Bhairav. Lāṭ Bhairav Tālāba now called as Kapālmocana	Condition: Restoration work	 May 2019
28.	Karṇaghaṇṭā	Ghaṇṭākarṇa Hrada	Karṇaghaṇṭā	 May 2019
29.	Khajuā Tālāba			Extinct
30.	Kṛmi Kuṇḍa	Kṛmi Kuṇḍa, Mahāsiddhīśvar Kuṇḍa		 February 2017
31.	Kumun Gaḍḍhā			Extinct
32.	Kurukṣetra	Samnihatya, Kurukṣetra Kuṇḍa	Kurukśetra Pokharā (private)	 May 2019
33.	Lakṣmī Kuṇḍa	Mahālakśmī, Lakṣmī Kuṇḍa, Sri Kuṇḍa	Condition: Restoration Work, polluted	 May 2019
34.	Lolārka Kuṇḍa	Lolārka Tīrtha	Lolārka Kuṇḍa	 February 2017

No.	Names mentioned in the James Prinsep Map of 1822	Names from spatial texts and other scriptures	Present name and notes	Photograph
35.	Mahsaid Gaḍḍhā			Extinct
36.	Maladahiyā Tālāba			Extinct
37.	Mandākinī Tālāba	Mandākinī Tīrtha	Company Bagh Garden, Maidāgin Pokharā, Public Garden, Shahid Udyan	 May 2019
38.	Mānasarovara Kuṇḍa	Mānasarovara Kuṇḍa		 February 2016
39.	Mātṛ Kuṇḍa	Mātṛ Kuṇḍa	Mātṛ Kuṇḍa Condition: Restoration work	 May 2019
40.	Matsyodarī Tālāba	Matsyodarī	Machhodarī Park, Gokulchand Udyan Condition: Restoration work	 May 2019
41.	Misir Pokharā	Miśraka Tırtha, Dhruva Kuṇḍa	Misir Pokharā	Extinct
42.	Motī Jhīla			Extinct
43.	Murghābī Tālāba			Extinct
44.	Nāgotrī Gaḍḍhā			Extinct
45.	Nayāpokharī			Extinct
46.	Nodarī Gaḍḍhā			Extinct
47.	Omakāreśvar	Tāra Tīrtha Kuṇḍa, (Mahādeva Talāba)		

No.	Names mentioned in the James Prinsep Map of 1822	Names from spatial texts and other scriptures	Present name and notes	Photograph
48.	Pāpamōcana Kuṇḍa	Pāpamōcana Tīrtha		 February 2016 Extinct
49.	Piśācśamocan Kuṇḍa	Vimlodak Kuṇḍa	Piśhācamocan Kuṇḍa	 May 2019
50.	Pitṛ Kuṇḍa	Pitṛ Kuṇḍa	Pitara Kuṇḍa Condition: Restoration work	 May 2019
51.	Puṣkara Talāba	Ayogandh Kuṇḍa	Puṣkara Tālāba Condition: Restoration work	 May 2019
52.	Rāma Kaṭorā Tālāba		Rāma Kaṭorā	 August 2018
53.	Rāma Kuṇḍa	Rāma Kuṇḍa	Rāma Kuṇḍa	 August 2018

No.	Names mentioned in the James Prinsep Map of 1822	Names from spatial texts and other scriptures	Present name and notes	Photograph
54.	Rānī Bhavānī Talāba	Kapālmocana Tīrtha-Sarovara	Rānī Bhavānī restored this Talāba	Extinct
55.	Revḍī Tālāba	Revā Tīrtha		Extinct
56.	Ṛṇamocana Tālāba (Ludhoo … Talāba)	Ṛṇamocana Tīrtha	Luddhu Gaḍḍhā	Image is not available
57.	Sahoṭiyā Talāba	Bhadravan Tīrtha	Sonabhadra Talāba	 August 2018
58.	Śaṅkūdhārā Talāba	Śaṅkhodvār Tīrtha	Śaṅkūdhārā Kuṇḍa, Dwarikadhish Mandir	 August 2018
59.	Śhaunak kuṇḍa	Śhaunak Kuṇḍa		Extinct
60.	Sonariā Tālāba			Extinct
61.	Sūraja Kuṇḍa	Sūrya Kuṇḍa, Sāṃbāditya Kuṇḍa		 August 2018
62.	Umariyā Tālāba	Near Lāṭ Bhairav (probably Lath Saraiya)		Image is not available

Source: Data collected by the author from 2016-19. All photographs are by the author.

APPENDIX IV

Population of Varanasi City (during the Colonial Period)

Year	Population
Zulfiqar Ali Khan Estimate (1801)	582,625
James Prinsep Survey in 1829	181,482
A survey by A.P. Currie in 1837	173,097
A survey by Tucker in 1852	185,984
A survey by Gubbins in 1853	181,330
Census of 1881	218,573
Census of 1891	223,375
Census of 1901	209,331
Census of 1911	205,420
Census of 1921	200,022
Census of 1931	207,650
Census of 1941	266,002

Sources: Valentia 1811; Greaves 1909; Medhasananda 2002; Singh 2009.

Bibliography

PRIMARY SOURCES

Aiyangar, Rangaswami K.V. (ed.), 1942. *Kṛtyakalpataru of Bhatta Lakṣmīdhara*, Baroda: Oriental Institute.

Apte, Vaman Shivaram 1957-9, Revised and Enlarged Edition of Prin. V.S. Apte's *The Practical Sanskrit-English Dictionary*, Poona: Prasad Prakashan, (https://dsal.uchicago.edu/dictionaries/apte/)

Elliott, Julia (ed.), 1997 *The Oxford Compact Dictionary & Thesaurus*, Oxford University Press.

Gangadharan, N. (tr. and annotated), (first edition in 1954), 1984, *The Agni-Purāṇa*, Delhi: Motilal Banarsidass.

Joshi, K.L., 2007, *Matsya Mahāpurāṇa*, vol. 1, Delhi: Parimal Publications.

Kemp, P.M. (tr. and ed.), 1959, *Russian Travellers to India and Persia (1624-1798) Kotov-Yefremov-Danibegov*, Delhi: Jiwan Prakashan.

Khandelwal, S.N. (tr.), 2015, *Skanda Maha Puranam of Vedvyasa: Kashi Khanda*, Varanasi: Chowkhamba Sanskrit Series.

Krishnacharya, T.R. and T.R. Vyasacharya, 1908, *Srimanmahabharatam: A new Edition Mainly Based on The South Indian texts, with Footnotes and Readings*, Vanaparva III, Bombay: Javaji Dadaji's Nirnayasagar Press.

Merriam-Webster's Dictionary of Synonyms, 1984, Merriam-Webster, Inc.

Molesworth, James Thomas, 1857, *A Dictionary, Marathi and English*, (second edition), Bombay: Bombay Education Society's press.

Sardesai, Govind Sakharam, 1948, *New History of The Marathas: The Expansion of the Maratha Power 1707-72*, vol. 2, Bombay: K.B. Dhawale for Phoenix Publications.

Shastri, J.L. (ed.), 1973, *The Liṅga-Purāṇa*, Delhi: Motilal Banarsidass.

——, 1985, *The Brahma-Purāṇa*, Delhi: Motilal Banarsidass.

——, 1992, *The Garuḍa-Purāṇa*, Delhi: Motilal Banarsidass.

Smrititirtha, Kamalakrishna (ed.), 1912, *Tīrthacintāmaṇi of Vācaspati Miśra*, Calcutta: Sanskrit Press.

Sukthankar, Vishnu S., 1942, *The Mahābhārata: The Āraṇyakaparvan*, Poona: Bhandarkar Oriental Research Institute.

Sukthankar, Vishnu S. et al., 1919-66, *The Critical Edition of Mahabharata*, Pune: Bhandarkar Oriental Research Institute.

Tagare, Ganesh Vasudeo (tr. and annot.), 1980, *The Nārada Purāṇa*, Delhi: Motilal Banarsidass.

——, 1981, *The Kūrma Purāṇa*, Delhi: Motilal Banarsidass.

——, 1987-8, *The Vāyu Purāṇa*, Delhi: Motilal Banarsidass.

——, 2007, (first edition 1992), *The Skanda Purāṇa*, Part I, Delhi: Motilal Banarsidass.

——, 1996-7, *Kāśī Khaṇḍa, Skanda Purāṇa*, Delhi: Motilal Banarsidass.

PUBLISHED REPORTS AND RECORDS

Benares and its Ghats, 1931, Kashi Tirtha Sudhar Trust, Benares, Allahabad: Allahabad Law Journal Press.

Blunt, E.A.H., 1912, *Census of India*, 1911, vol. XV: *United Provinces of Agra and Oudh*, Part 1, Report, Allahabad: Superintendent, Government Press.

Buckland, C.E., 1906, *Dictionary of Indian Biography*, London: Swan Sonnenschein and Co.

Calendar of Persian Correspondence: Being Letters, Referring Mainly to Affairs in Bengal, Which Passed Between Some of the Company's Servants and Indian Rulers and Notables, (1792-3), 1959, vol. X, Delhi: National Archives of India.

Chadwick, Edwin, 1842, *Report to Her Majesty's Principal Secretary of State for the Home Department from the Poor Law Commissioners, on an Inquiry into the Sanitary Condition of the Labouring Population of Great Britain; with Appendices.* (Presented to both Houses of Parliament, by Command of Her Majesty, July 1842), London: W. Clowes and Sons.

Hart, Ernest (ed.), 1889, 'India and the Colonies', *The British Medical Journal*, vol. 1, London: The British Medical Association.

Report from the Select Committee on the Affairs of the East India Company: With Minutes of Evidence in Six Parts and an Appendix and Index to Each, 1832, House of Commons, Great Britain.

Report of the Commissioners Appointed to Inquire into the Sanitary State of the Army in India; with Abstract of Evidence, and of Reports received from Indian Military Stations, August 1864, London: Superintendence of Her Majesty's Stationery Office.

Supplement to the Indian Journal of Medical Research, Proceedings of the Third All-India Sanitary Conference held at Lucknow, (19-27 January 1914), vol. I: *Discussion and Resolutions*, Calcutta: Thacker Spink & Co.

The House of Commons, 1803, *Journals of the House of Commons (From 24 January 1786 to 14 December 1786, King George the Third*), vol. 41.

The Imperial Gazetteer of India, Provincial Series, *United Provinces of Agra and Oudh*, vol. I and II, *The Provinces; Rivers, Mountains, Lakes, Canals, and Historic Areas; The Meerut, Agra and Bareilly Divisions*, 1908, Calcutta: Superintendent of Government Printing.

The Imperial Gazetteer of India, Provincial Series, *United Provinces of Agra and Oudh*, vol. II: *The Provinces; Rivers, Mountains, Lakes, Canals, and Historic Areas; The Meerut, Agra and Bareilly Divisions*, 1908, Calcutta: Superintendent of Government Printing.

SECONDARY SOURCES

Agarwal, Anil and Sunita Narain, (eds), 1997, *Dying Wisdom: Rise, Fall and Potential of India's Traditional Water Harvesting Systems*. New Delhi: Centre for Science and Environment.

Agrawala, Vasudeva Sharana, 1953, *India as Known to Pāṇini: A Study of the Cultural Material in the Aṣṭādhyāyī*, Lucknow: University of Lucknow.

Aiyar, Narayanswami K., 1914, *The Puranas: In the Light of Modern Science*, Madras: Theosophical Society.

Alberti, Marina., John. M. Marzluff, Eric Shulenberger, Gordon Bradley, Clare Ryan and Craig Zumbrunnen, 2003, 'Integrating Humans into Ecology: Opportunities and Challenges for Studying Urban Ecosystems', *BioScience*, 53.12: 1169-79.

Ali, S.M., 1966, *The Geography of the Puranas*, New Delhi: People's Publishing House.

Altekar, Anant Sadashiv, 1947, *Benares and Sarnath: Past and Present*, Varanasi: Banaras Hindu University.

'An Old Resident', 1917, *All About Benares: Containing A Sketch from the Vedic Days to the Modern Times*, Benares City: K.S. Muthiah & Co.

Arnold, David, 2000, *The New Cambridge History of India, Science, Technology and Medicine in Colonial India*, Cambridge: Cambridge University Press.

——, 1993, *Colonizing the Body: State Medicine and Epidemic Disease in Nineteenth-Century India*, California: University of California Press.

——, 1989, 'The Ecology and Cosmology of Disease in the Banaras Region, in Sandria B. Freitag (ed.) *Culture and Power in Banaras. Community, Performance and Environment, 1800-1980*, Berkeley: University of California Press, pp. 246-67.

Awasthi, A.B.L., 1965, *Studies in Skanda Purana*, Lucknow: Kailash Prakashan.

Bakker, Hans, 2009, 'Rama Devotion in a Saiva Holy Place: The Case of Varanasi', in Maria Pauwels Heidi Rika. (ed.) *Patronage and Popularisation, Pilgrimage and Procession: Channels of Transcultural Translation and Transmission in*

Early Modern South Asia; Papers in Honour of Monika Horstmann, vol. 58 of Studies in Oriental Religions, Wiesbaden: Harrassowitz Verlag, pp. 67-79.

Bartolomeo, Paolino Da San, 1800, *A Voyage to the East Indies*, London: J. Davis.

Beal, Samuel, 1884, *Si-Yu-Ki, Buddhist Records of the Western World: Translated from the Chinese of Hiuen Tsiang (AD 629)*, vol. 2, London: Trübner and Co.

Benton-Short, Lisa and John Rennie Short, 2008, *Cities and Nature*, Oxford: Routledge Taylor & Francis Group.

Bhardwaj, Surinder M., 1983, *Hindu Places of Pilgrimage in India: A Study in Cultural Geography*, California: University of California Press.

Birch, T.W., 1949, *Maps: Topographical and Statistical*, London: Oxford University Press.

Bisschop, Peter, 2013, 'The Abode of the Pañcamudrās: A Yoginī Temple in Early Medieval Vārānasī' in István Keul (ed.), 2013, *Yogini in South Asia*, London: Routledge, pp. 47-60.

Blunt, E.A.H., 1912, *Census of India, 1911*, vol. XV: *United Provinces of Agra and Oudh*, Part 1: *Report*, Allahabad: F. Luker, Superintendent, Government Press.

Boyden, Stephen, Sheelagh Millar, Ken Newcombe and Beverley O'Neill, 1981, *The Ecology of a City and Its People: The Case of Hong Kong*, Canberra: Australian National University Press.

Broich, John, 2007, 'Engineering the Empire: British Water Supply Systems and Colonial Societies, 1850-1900', *Journal of British Studies*, 46.2: pp. 346-65.

Burton, Isabel, 1879, *Arabia Egypt India A Narrative of Travel*, London and Belfast: William Mullan and Son.

Buyers, William, 1848, *Recollections of Northern India*, London: John Snow Paternoster Row.

Candiani, Vera S., 2014, *Dreaming of Dry Land: Environmental Transformation in Colonial Mexico City*, California: Stanford University Press.

Cape, Charles Phillips, 1910, *Benares: The Stronghold of Hinduism*, Boston: R. G. Badger.

Caunter, Hobart, 1834, *The Oriental Annual, or Scenes in India*, London: Edward Bull.

Céréghino, Régis, Dani Boix, Henry-Michel Cauchie, Koen Martens and Beat Oertli, 2014, 'The Ecological Role of Ponds in a Changing World', *Hydrobiologia*, 723, pp. 1-6.

Cohn, Bernard S., 1960, 'The Initial British Impact on India: A Case Study of the Benares Region', *The Journal of Asian Studies*, 19.4: 418-31.

——, 1962, 'The British in Benares: A Nineteenth Century Colonial Society', *Comparative Studies in Society and History*, 4.2: 169-99.

Coleman, B. I., 2012, *The Idea of the City in Nineteenth-Century Britain*, London and Boston: Routledge & Kegan Paul.

Colvin, Auckland, 1894, 'Municipal and Village Water-Supply and Sanitation in the Northern West Provinces and Oudh', *The Journal of the Society of Arts*, 42.2164: 515-46, Royal Society for the Encouragement of Arts, Manufactures and Commerce.

Crooke, William (ed. second edition), 1925, *Travels in India by Jean-Baptiste Tavernier* (tr. by V. Ball), London: Oxford University Press.

Cunningham, Alexander, 1871, *Four Reports Made During the Years 1862-63-64-65*, vol. I, Simla: The Government Central Press.

Da Cunha, Dilip, Geeta Mehta, Kate Orff and Julia Watson, 2018, 'From Anandavana to Varanasi', *Water Urbanism*, New York: Columbia University. https://www.arch.columbia.edu/books/reader/331-water-urbanism-varanasi

Da Cunha, Dilip, 2019, *The Invention of Rivers: Alexander's Eye and Ganga's Descent*, Philadelphia: University of Pennsylvania Press.

Dar, Vrinda, 2005, 'Threats and Prospects', in George Michell and Rana P. B. Singh (eds.) *Banaras The City Revealed*, Mumbai: Marg Publications. pp. 139-43.

Davies, B.R., J. Biggs, P.J. Williams, J.T. Lee and S. Thompson, 2007, 'A Comparison of the Catchment Sizes of Rivers, Streams, Ponds, Ditches and Lakes: Implications for Protecting Aquatic Biodiversity in an Agricultural Landscape', in Beat Oertli, Régis Céréghino, Jeremy Biggs, Steven Declerck, Andrew Hull and Maria Rosa Miracle (eds.) *Pond Conservation in Europe: Developments in Hydrobiology*, 210, Dordrecht: Springer, pp. 7-17.

Denney, Joseph Villiers (ed.), 1907, *Macaulay's Essay on Warren Hastings*, Boston: Allyn and bacon.

Devasthali, G.V. Sanskrit, in R.C. Majumdar, (ed.), 1954, *The History and Culture of the Indian People: The Classical Age*, Bombay: Bharatiya Vidya Bhavan, pp. 291-324.

D'Souza, Rohan, 2006, 'Water in British India: The Making of a "Colonial Hydrology"', *History Compass* 4.4: 621-8.

——, 2015, 'Drainage, River Erosion and Chaurs: An Environmental History of Land in Colonial Eastern India', *Working Paper Series. History and Society* 84, Nehru Memorial Museum and Library, pp. 3-23.

D'Souza, Rohan and Harini Nagendra, 2011, 'Changes in Public Commons as a Consequence of Urbanization: The Agara Lake in Bangalore, India', *Environmental Management*, 47, 840-50.

Duncan, Jonathan, 1799, 'An Account of the Discovery of Two Urns in the Vicinity of Benares', in *Asiatic Researches; or, Transactions of the Society Instituted in Bengal, for Inquiring into the History and Antiquities, the Arts, Sciences, and Literature of Asia*, London, pp. 131-2.

Eck, Diana, L., 1981, 'India's "Tirthas": "Crossings" in *Sacred Geography. History of Religions*' 20.4: 323-44.

——, 1985, 'Banaras: Cosmos and Paradise in the Hindu Imagination' *Contributions to Indian Sociology*, 19.1: 41-55.

——, 2012, *India: A Sacred Geography*, New York: Three Rivers Press.

——, 2015 (first edition in 1983), *Banaras: City of Light*, New Delhi: Penguin India.

Edmondson, Beth and Stuart Levy, 2013, *Climate Change and Order: The End of Prosperity and Democracy*, London: Palgrave Macmillan.

Edney, Matthew H., 1997, *Mapping an Empire: The Geographical Construction of British India, 1765-1843*, Chicago and London: The University of Chicago Press.

Fisher, F.H. and J.P. Hewett, 1884, *Statistical, Descriptive, and Historical Account of the North-Western Provinces of India*, vol. XIV, Part I: *Benares*, Allahabad: North-Western Provinces and Oudh Government Press.

Fitter, R.S.R., 1945, *London's Natural History*, London: Readers Union Collins.

Forman, Richard, T.T., 2014, *Urban Ecology Science of Cities*, Cambridge: Cambridge University Press.

Gandy, Matthew, 2013, 'Marginalia: Aesthetics, Ecology, and Urban Wastelands', *Annals of the Association of American Geographers*, 103.6: 1301-16.

Gengnagel, Jörg, 2011, *Visualized Texts: Sacred Spaces, Spatial Texts and the Religious Cartography of Banaras*, Wiesbaden: Harrassowitz Verlag.

Giles, H.A., 1923, *The Travels of Fa-Hsien (AD 399-414), or Record of the Buddhistic Kingdoms*, Cambridge: University Press.

Gogate, Prasad P. and B. Arunachalam, 1998, 'Area Maps in Maratha Cartography: A Study in Native Maps of Western India', *Imago Mundi*, 50.1: 126-40.

Gooley, Tristan, 2016, *How to Read Water: Clues and Patterns from Puddles to the Sea*, London: Sceptre.

Greaves, Edwin, 1909, *Kashi: The City Illustrious or, Benares*, Allahabad: The Indian Press.

Grimm, N.B., J.M. Grove, S.T.A. Pickett and C.A. Redman, 2000, 'Integrated Approaches to Long-Term Studies of Urban Ecological Systems', *Bio Science*, 50: 571-84.

Grove, Morgan J., Mary L. Cadenasso, Steward T. A. Pickett, Gary E. Machlis and William R. Burch Jr., 2015, *The Baltimore School of Urban Ecology:*

Space, Scale, and Time for the Study of Cities, London: Yale University Press.

Gupta, Anil Kumar and Sreeja S. Nair, 2011, Urban Floods in Bangalore and Chennai: Risk Management Challenges and Lessons for Sustainable Urban Ecology', *Current Science*, 100: 1638-45.

Gupta, Kapil, 2005, 'The Drainage Systems of India's Cities', *Waterlines*, 23.4: 22-4.

Gutschow, Niels, 2006, *Benares: The Sacred Landscape of Varanasi*, Stuttgart and London: Edition Axel Menges.

Hanley, James, 2002, 'Edwin Chadwick and the Poverty of Statistics', *Medical History*, 46: 21-40.

Harrison, Mark, 2011, 'Medicine and Colonialism in South Asia Since 1500', in Mark Jackson (ed.) *The Oxford Handbook of the History of Medicine*, Oxford University Press, pp. 285-301.

Hassall, Christopher, 2014, 'The Ecology and Biodiversity of Urban Ponds', *WIREs Water*, 1: 187-206.

Hazra, Rajendra Chandra, 1940, *Studies in the Purānic Records on Hindu Rites and Customs*, Dacca: The University of Dacca.

Heber, Reginald, 1828, *Narrative of a Journey through the Upper Provinces of India: From Calcutta to Bombay, 1824-5, (with notes upon Ceylon), an Account of a Journey to Madras and the Southern Provinces, 1826, and Letters Written in India*, vol. II, London: John Murray.

Hegewald, Julia A.B., 2002, *Water Architecture in South Asia: A Study of Types, Developments and Meanings*, Leiden: Brill.

Hodges, William, 1793, *Travels in India, During the years 1780, 1781, 1782, & 1783*, London: J. Edwards, Pall-Mall.

Hubert, Jane, 1994, 'Sacred Beliefs and Beliefs of Sacredness', in L. David, Jane Hubert Carmichael, Brian Reeves and Audhild Schanche (ed.) *Sacred Sites, Sacred Places*, London: Routledge, pp. 9-19.

Hume, John Chandler, 1986, 'Colonialism and Sanitary Medicine: The Development of Preventive Health Policy in the Punjab, 1860 to 1900', *Modern Asian Studies*, 20.4: 703-24.

Hurst, Fletcher John, 1891, *Indika: The Country and the People of India and Ceylon*, New York: Harper & Brothers, Franklin Square.

Irvine, William (tr.), 1913, *A Pepys of Mogul India: 1653-1708–Being an abridged Edition of the 'Storia Do Mogor' of Niccolao Manucci*, New York: E.P. Dutton and Company.

Jacobsen, Knut A., 2013, *Pilgrimage in the Hindu Tradition: Salvific Space*, London: Routledge.

Jarrett, H.S., (tr.), 1948, *'Ain-i-Akbari of Abul Fazl-i-'Allami*, vol. III, Calcutta:

Royal Asiatic Society of Bengal, (revised and further annotated by Jadunath Sarkar).

——, 1949, (second edition-corrected and annotated by Jadunath Sarkar), 1949, *Ain-i-Ākbari of Abul Fazl-i-Allami*, vol. II, Calcutta: Royal Asiatic Society of Bengal.

Jayaswal, Vidula and Uma Kant Shukla, 2016, Summary and Deductions, in Vidula Jayaswal, Uma Kant Shukla, and Meera Sharma (eds.) *Varanasi and Ganga: Geomorphological and Archaeological Investigations at Ramnagar*, New Delhi: Aryan Books International, pp. 317-31.

Jayaswal, Vidula. 2015, 'Sacred Spaces of the Middle Ganga Valley: A Case Study of Varanasi', in Himanshu Prabha Ray (ed.) *Negotiating Cultural Identity: Landscapes in Early Medieval South Asian History*, New Delhi: Routledge.

Jayaswal, Vidula, Manoj Kumar, 2005-6, 'Excavations at Ramnagar: Discovery of a Supporting Settlement of Ancient Varanasi', in K.N. Dikshit (ed.) *Puratattva*, 36, New Delhi: Indian Archaeological Society, pp. 85-92.

Kane, Pandurang Vaman, 1953, *History of Dharmasastra*, vol. IV, Poona: Bhandarkar Oriental Research Institute.

Kantawala, S.G., 1995, 'Puranas on Pilgrimage: Further Remarks', in D.P. Dubey (ed.) *Pilgrimage Studies: Sacred Places, Sacred Traditions*, Allahabad: The Society of Pilgrimage Studies, pp. 25-30.

Kennedy, James, 1884, *Life and Work in Benares and Kumaon, 1839-77*, London: T. Fisher Unwin.

Kishtawal, C.M., Dev Niyogi, Mukul Tewari, Roger A. Pielke Sr. and Marshall J. Shephard, 2009, 'Urbanization Signature in the Observed Heavy Rainfall Climatology Over India, *International Journal of Climatology*, 30: 1908-16.

Kopf, David, 1969, *British Orientalism and the Bengal Renaissance: The Dynamics of Indian Modernization 1773-1835*, Calcutta: Firma K.L. Mukhopadhyay.

Kundu, Debolina and Dibyendu Samanta, 2011, 'Redefining the Inclusive Urban Agenda in India', *Economic & Political Weekly*, XLV.5: 55-63.

Lannoy, Richard, 2001-2, Benares as Tirtha, 'India International Centre Quarterly', 28.4: 376-86.

Law, Bimala Churn, 1932, *Geography of Early Buddhism*, London: Kegan Paul, Trench Trübner and Co.

——, 1976, *Historical Geography of Ancient India*, Delhi: Ess Ess Publications.

Lazzaretti, Vera, 2017, 'Waterscapes in Transition: Past and Present Reshaping of Sacred Water Places in Banaras, in Ravi Baghel, Lea Stepan and Joseph K.W. Hill (eds.) *Water, Knowledge and the Environment in Asia: Epistemo-*

logies, Practices and Locales, London and New York: Routledge, pp. 230-45.

Lee, Jackson, 2014, *Dirty Old London: The Victorian Fight Against Filth*, London: Yale University Press.

Legge, James (tr. and annot.), 2003 (first published in 1886), *A Record of Buddhistic Kingdoms, Being an Account by The Chinese Monk Fa-Hien of His Travels in India and Ceylon (AD 399-414) in Search of the Buddhist Books of Discipline*. Delhi: Asian Educational Services.

Mahomet, Sake Dean, 1794, *The Travels of Dean Mahomet: A Native of Patna in Bengal, through Several Parts of India, while in the Service of the Honourable the East India Company*, vol. 1, Cork: J. Connor.

Mani, Vettam, 1975, *Puranic Encyclopaedia*, Delhi: Motilal Banarsidass.

Marcotullio, Peter J. and William Solecki, 2013, 'What Is a City?: An Essential Definition for Sustainability', in Christopher G. Boone and Michail Fragkias (eds.) *Urbanization and Sustainability: Linking Urban Ecology, Environmental Justice and Global Environmental Change*, Dordrecht: Springer Science + Business Media, pp. 11-25.

Martin, Renald J., 1868, *The Sanitary History of the British Army in India: Past and Present*, London: Savill, Edwards and Co.

McDonald, Robert I., Peter J. Marcotullio and Burak Güneralp, 2013, 'Urbanization and Global Trends in Biodiversity and Ecosystem Services', in Thomas Elmqvist, Michail Fragkias, Julie Goodness, Burak Güneralp, Peter J. Marcotullio, Robert I. McDonald, Susan Parnell, Maria Schewenius, Marte Sendstad, Karen C. Seto and Cathy Wilkinson (eds.) *Urbanization, Biodiversity and Ecosystem Services: Challenges and Opportunities*, Dordrecht Heidelberg New York London: Springer, pp. 31-52.

McDonnell, Mark, J., 2011, 'The History of Urban Ecology: An Ecologist's Perspective', in Jari Niemelä, Jürgen H. Breuste, Thomas Elmqvist, Glenn Guntenspergen, Philip James and Nancy E. McIntyre (eds.) *Urban Ecology: Patterns, Processes, and Applications*. Oxford: Oxford University Press, pp. 5-13.

Medhasananda, Swami, 2010 (first edition 2002), *Varanasi at the Crossroads: A Panoramic View of Early Modern Varanasi and the Story of Its Transition* (2 volumes), Kolkata: Ramakrishna Mission, Institute of Culture.

Michaels, Axel, 2006, 'Mapping the Religious and Religious Maps: Aspects of Transcendence and Translocality in Two Maps of Varanasi', in Martin Gaenszle and Jörg Gengnagel (eds.) *Visualised Space in Banaras: Images, Maps, and the Practice of Representation*, Wiesbaden: Harrassowitz Verlag, pp. 131-43.

Mohan, Kshitij, Ajai Srivastava and Kumar Praveen Rai, 2011, 'Ground Water in the City Varanasi, India: Present Status and Prospects', *Quaestiones Geographcae*, 30.3: 47-60.

Mohan, Rakesh., and Shubhagato Dasgupta, 2005, 'The 21st Century: Asia Becomes Urban', *Economic and Political Weekly*, 40.3: 213-23.

Murdoch, John, 1897, *Kasi, or Benares: The Holy City of the Hindus*, London and Madras: The Christian Literature Society for India.

Muthiah, K.S., 1911, *Smiling Benares: Containing A Sketch from the Vedic Days to the Modern Times*, Madras: Raithby & Co.

Nagendra, Harini, H.S. Sudhira, Madhusudan Katti and Maria Schewenius, 2013, 'Sub-regional Assessment of India: Effects of Urbanization on Land Use, Biodiversity and Ecosystem Services', in Thomas Elmqvist, Michail Fragkias, Julie Goodness, Burak Güneralp, Peter J. Marcotullio, Robert I. McDonald, Susan Parnell, Maria Schewenius, Marte. Sendstad, Karen C. Seto and Cathy Wilkinson, (eds.) *Urbanization, Biodiversity and Ecosystem Services: Challenges and Opportunities: A Global Assessment*, pp. 75-91.

Narain, Sunita, 2016, 'An urban Nightmare', in Rajit Sengupta, (eds.), 'Why Urban India Floods. Indian Cities Grow at the Cost of their Wetlands', *Down to Earth*, New Delhi: Centre for Science and Environment, pp. 4-5.

Narain, Vishnu, Anugrah, 1959, *Jonathan Duncan and Varanasi*, Calcutta: K. L. Mukhopadhyay.

Narasimhan, Balaji. S. Murty Bhallamudi, Arpita Mondal, Subimal Ghosh and Pradeep Mujumdar, 2016, *Chennai Floods 2015: A Rapid Assessment*. Bangalore: Interdisciplinary Centre for Water Research Indian Institute of Science.

Nevill, H.R., 1909, *Benares: A Gazetteer, vol. XXVI of the District Gazetteers of the United Provinces of Agra and Oudh*, Allahabad: Government Press.

Niyogi, Roma, 1959, *The History of the Gahadavala Dynasty*, Calcutta: Oriental Book Agency.

Oldenburg, Veena Talwar, 1984, *The Making of Colonial Lucknow, 1856-1877*, New Jersey: Princeton University Press.

Oldham, Wilton, 1876, *North Western Provinces. Historical and Statistical Memoir of the Ghazeepoor District*, Part II: *History of Ghazeepoor and the Benares Province from AD 1781 to AD 1795*, Allahabad: North-Western Provinces Government Press.

Pandey, Raj Bali, 1949, 'The Historical Interpretation of the Nadi-Stuti Hymn in the Ṛgveda', *Proceedings of the Indian History Congress*, 12: 93-6.

Parker, Arthur, 1901, *A Handbook of Benares*, (second edition), Tribandrum: Travancore Government Press.

Pataki, Diane E., 2015, 'Grand Challenges in Urban Ecology', *Frontiers in Ecology and Evolution*, 3.57: 1-6.

Phillimore, R.H. (collected and compiled), 1945, *Historical Records of the Survey of India*, vol. 1, *18th Century*, Dehradun: Survey of India.

Phillimore, R.H. 1954, *Historical Records of the Survey of India*, vol. III: *1815-1830*, Dehradun: Office of the Geodetic Branch Survey of India.

Pickett, Steward T.A., Mary. L. Cadenasso, Daniel L. Childers, Mark J. McDonnell and Weiqi Zhou, 2016, 'Evolution and Future of Urban Ecological Science: Ecology in, of, and for the City', *Ecosystem Health and Sustainability*, 2.7: 1-16.

Pieper, Jan. 1979, 'A Pilgrim's Map of Benares', The Asian Monsoon and its Economic Consequences. *GeoJournal*, 3.2: 215-18.

Prinsep, James, 1831, *Benares Illustrated, in a Series of Drawings*, Calcutta: The Baptist Mission Press.

——, 1832, 'Census of the Population of the City of Benares', in *Asiatic Researches; or Transactions of the Society*, vol. XVII, Calcutta: Bengal Military Orphan Press, pp. 470-98.

Pusalker, A.D., 1955, *Studies in the Epics and Purāṇas*, Bombay: Bharatiya Vidya Bhavan.

Raj, Kapil, 2007, *Relocating Modern Science: Circulation and the Construction of Knowledge in South Asia and Europe, 1650-1900*, Palgrave Macmillan.

Raju, Prudhvi K.N. and Manish Kumar Pandey, 2013, 'Varanasi: Origin and Growth from a Geomorphic Perspective', in Vidula Jayaswal (ed.) *Varanasi: Myths and Scientific Studies*, New Delhi: Aryan Books International, pp.134-48.

Raju, Prudhvi K.N. and Diva Bhatt, 2015, 'Water in Ancient Indian Perspective and Ponds of Varanasi as Water Harvesting Structures', in Janardhan N. Raju, Wolfgang Gossel, A.L. Ramanathan and M. Sudhakar (eds.) *Management of Water, Energy and Bio-resources in the Era of Climate Change: Emerging Issues and Challenges*. Springer International Publishing, pp. 63-71.

Rapson, E.J. (ed.), 1922, *Cambridge History of India*, vol. 1: *Ancient India*, Cambridge University Press.

Rennell, James, 1793, *Memoir of a Map of Hindoostan or the Mogul Empire*, (third edition), London: W. Bulmer and Co.

Reynolds-Ball, Eustace, 1907, *The Tourist's India*, London: Swan Sonnenschein & Co.

Richardson, Benjamin Ward, 1887, *The Health of Nations: A Review of the Works of Edwin Chadwick with a Biographical Dissertation*, vol. I, London: Longmans, Green and Co.

Rocher, Ludo, 1986, *The Purāṇas* (A History of Indian Literature, vol. 2, Fasc. 3), Wiesbaden: Harrassowitz.

Rötzer, Klaus, 2005, 'Wells, Tanks, and Gardens', in George Michell and Rana P. B. Singh (eds.) *Banaras: The City Revealed*, Mumbai: Marg Publications, pp. 117-25.

Ryley, Horton J., 1899, *Ralph Fitch: England's Pioneer to India and Burma*, London: T. Fisher Unwin.

Sachau, Edward, 1888, *Alberuni's India: An Account of the Religion, Philosophy, Literature, Geography, Chronology, Astronomy, Customs, Laws and Astrology of India About AD 1030*, vol. II, London: Trübner & Co.

Sandes, E.W.C., 1933-5, *The Military Engineer in India*, vol. I & II, Chatham: The Institution of Royal Engineers.

Saraswati, Baidyanath, 1975, *Kashi: Myth and Reality of a Classical Cultural Tradition*, Simla: Indian Institute of Advanced Study.

——, 1984, *The Spectrum of the Sacred: Essays on the Religious Traditions of India*, New Delhi: Concept Publishing Company.

Sastri, Ramaswami K.S., 1937, 'The Puranas and their Contribution to Indian Thought', in *The Cultural Heritage of India*, vol. 1, Calcutta: Sri Ramakrishna Centenary Committee, pp. 169-82.

Sastri, Subrahmanya V. and Ramakrislina M. Bhat, 1947, *Varahamihira's Brihat Samhita*, Bangalore: V.B. Soobbiah & Sons, 1947.

Schwartzberg, Joseph E., 1992, 'Introduction to South Asian Cartography', in J.B. Harley and David Woodward (eds.) *Cartography in the Traditional Islamic and South Asian Societies: The History of Cartography*, vol. 2, Book 1, Chicago: The University of Chicago Press.

Seto, Karen C., Susan Parnell and Thomas Elmqvist, 2013, 'A Global Outlook on Urbanization', in Thomas Elmqvist, Michail Fragkias, Julie Goodness, Burak Güneralp, Peter J. Marcotullio, Robert I. McDonald, , Susan Parnell, Maria Schewenius, Marte Sendstad, Karen C. Seto and Cathy Wilkinson (eds.) *Urbanization, Biodiversity and Ecosystem Services: Challenges and Opportunities*, Dordrecht Heidelberg New York London: Springer, pp. 1-12.

Shakespear, Alexander, 1873, *Selections from the Duncan Records*, vol. 1, Benares: The Medical Hall Press.

Shastri, Haraprasada, 1928, *A Descriptive Catalogue of the Sanskrit Manuscripts in the Government Collection Under the Care of The Asiatic Society of Bengal*, vol. V: *Purana Manuscript*, Calcutta: Asiatic Society of Bengal.

Sherring, Matthew Atmore, 1868, *The Sacred City of the Hindus: An Account of Benares in Ancient and Modern Times*, London: Trübner & Co.

Shukla, Uma Kant, Pradeep Srivastava and I.B. Singh, 2012, 'Migration of the Ganga River and Development of Cliffs in the Varanasi Region, India During the Late Quaternary: Role of Active Tectonics', *Geomorphology*, 171-2, pp. 101-13.

Singh, Rana P.B., 2016, 'Urban Heritage and Planning in India: A Study of Banaras, in Ashok K. Dutt, Allen G. Noble, Frank G. Costa, Rajiv R. Thakur and S.K. Thakur (eds.) *Spatial Diversity and Dynamics in Resources and Urban Development*, vol. II, Urban Development: Dordrecht, New York: Springer Science+Business Media, pp. 423-49.

——, 1994, 'Water Symbolism and Sacred Landscape in Hinduism: A Study of Benares (Vārāṇasī) [Wassersymbolismus und heilige Landschaft im Hinduismus: Eine Studie aus Benares (Vārāṇasī)]', *Erdkunde*, 48.3: 210-27.

——, 1993, 'Varanasi, A World Heritage City: Frame, and Historical Accounts', in Rana P.B. Singh, (eds.) *Banaras (Varanasi), Cosmic Order, Sacred City, Hindu Traditions*, Varanasi: Tara Book Agency, pp. 297-307.

Singh, R.L., 1955, *Banaras: A Study in Urban Geography*, Banaras: Nand Kishore & Sons.

Singh, S.P., 1995, 'Origin and Growth of the Institution of Pilgrimage', in D.P. Dubey (ed.) *Pilgrimage Studies: Sacred Places, Sacred Traditions*, Allahabad: The Society of Pilgrimage Studies.

Singh, U., 1999, 'Geographical Personality of Ganga Basin', in Virendra Kumar. Shrivastava, P.R. Chauhan and S.K. Dixit (eds.) *Commercial Activities and Development in the Ganga Basin*, New Delhi: Concept Publishing House, pp. 57-65.

Singh, Upinder, 2008, *A History of Ancient and Early Medieval India: From the Stone Age to the 12th Century*, India: Pearson Education.

Singh, Vijay P., Basant Maheshwari and Bhadranie Thoradeniya, 2016, 'Options and Strategies for Balanced Development for Liveable Cities: An Epilogue', in Basant Maheshwari, Vijay P. Singh, Thoradeniya Bhadranie (eds.) *Balanced Urban Development: Options and Strategies for Liveable Cities*, Water Science and Technology Library, vol. 72, Springer, pp. 589-601.

Sinha, Amita, 2020, 'Ghats on the Ganga in Varanasi, India: The Making of a Vernacular urban landscape and its Conservation', in Kapila. D. Silva (ed.) *The Routledge Handbook on Historic Urban Landscapes in the Asia-Pacific*, New York: Routledge, pp. 248-60.

——, 2011, *Landscapes in India: Forms and Meanings*, Delhi: Asian Educational Services.

Sinha, Surajit and Baidyanath Saraswati, 1978, *Ascetics of Kashi: An Anthropological Exploration*, Varanasi: N.K. Bose Memorial Foundation.

Sircar, Dineschandra, 1971, *Studies in the Religious Life of Ancient and Medieval India*, Delhi: Motilal Banarsidass.

Smith, Travis L., 2014, 'Vārāṇasī Envisioned: Approaches to the Purāṇic Māhātmyas', in István Keul (ed.) *Banāras Revisited: Scholarly Pilgrimages to the City of Light*, Wiesbaden: Harrassowitz Verlag, pp. 21-37.

Smith, Vincent A., (Second Edition) 1916, *Travels in the Mogul Empire: AD 1656-68,* by François Bernier and Archibald Constable (tr. and annot, 1891), Oxford University Press: Humphrey Milford.

Srinivasan, Veena, Steven M. Gorelick and Lawrence Goulder, 2010, 'Sustainable Urban Water Supply in South India: Desalination, Efficiency Improvement, or Rainwater Harvesting?', *Water Resources Research,* 46.10.

Sukopp, Herbert, 2002, 'On the Early History of Urban Ecology in Europe', *Preslia,* 74: 373-93.

——, 1998, 'Urban Ecology: Scientific and Practical Aspects', in Jürgen Breuste, Hildegard Feldmann and Ogarit Uhlmann (eds.) *Urban Ecology,* Berlin Heidelberg: Springer-Verlag, pp. 3-16.

Sukul, Kuber Nath, 1974, *Varanasi Down the Ages,* Patna: Kameshwar Nath Sukul.

Suryavanshi, Bhagwan Singh, 1986, *Geography of the Mahabharata,* New Delhi: Ramanand Vidya Bhawan.

Temple, Richard, 1898, *A Bird's-Eye View of Picturesque India,* London: Chatto & Windus.

Thoradeniya, Bhadranie and Malik Ranasinghe, 2016, 'Community Stakeholder Viewpoints on Issues of urbanisation Along the River Ma Oya, Sri Lanka', in Basant Maheshwari, Vijay P. Singh, Thoradeniya Bhadranie (eds.) *Balanced Urban Development: Options and Strategies for Liveable Cities,* Water Science and Technology Library, vol. 72, Springer, pp. 135-49.

Tiwari, Piyush., Ranesh Nair, Pavan Ankinapalli, Jyoti Rao, Pritika Hingorani and Manisha Gulati, 2015, *India's Reluctant Urbanization: Thinking Beyond,* Palgrave Macmillan.

Tjallingii Sybrand, 2012, 'Water Flows and Urban Planning', in Ellen van Bueren, Hein van Bohemen, Laure Itard, Henk Visscher (eds.) *Sustainable Urban Environments,* Dordrecht: Springer, pp. 91-111.

Tripathi, Maya Prasad, 1969, *Development of Geographic Knowledge in Ancient India,* Varanasi: Bharatiya Vidya Prakashan.

Tulchinsky, Theodore H. and Elena A. Varavikova, 2009, *The New Public Health: An Introduction for the 21st Century,* Elsevier Academic Press.

Twinning, William, H.G. (ed.), 1893, *Travels in India: A Hundred Years Ago with A Visit to the United States,* by *Thomas Twinning,* London: James R. Osgood, Msilvaine & Co.

Valdiya, K.S., 2012, *Geography, Peoples and Geodynamics of India in Puranas and Epics: A Geologist's Interpretations,* New Delhi: Aryan Books International.

Valentia, George Viscount, 1811, *Voyages and Travels to India, Ceylon, The Red Sea, Abyssinia, and Egypt, in the years 1802, 1803, 1804, 1805, and 1806*, London: F. C., and J. Rivington.

Van der Veer, Peter, 1988, *Gods on Earth: The Management of Religious Experience and Identity in a North Indian Pilgrimage Centre*, London: The Athlone Press.

Varady, Robert G., 1989, 'Land Use and Environmental Change in the Gangetic Plain: Nineteenth-Century Human Activity in the Banaras Region', in Sandria B. Freitag (ed.) *Culture and Power in Banaras: Community, Performance, and Environment, 1800-1980*, Berkeley: University of California Press, pp. 229-45.

Vidyarthi, Lalita Prasad, 1961 (second edition, 1978), *The Sacred Complex in Hindu Gaya*, Delhi: Concept Publishing Company.

Ward, Stephen V., 1994, *Planning and Urban Change*, London: Paul Chapman Publishing Ltd., p. 11.

Watson, Donald and Michaele Adams, 2011, *Design for Flooding: Architecture, Landscape, and Urban Design for Resilience to Flooding and Climate Change*, New Jersey: John Wiley & Sons, Inc.

Weiland, Ulrike and Matthais Richter, 2012. Urban Ecology, 'Brief History and Present Challenges', in Matthias Richter and Ulrike Weiland, (eds.), *Applied Urban Ecology: A Global Framework*, West Sussex, UK: Blackwell Publishing Ltd., pp. 3-11.

Wheeler, Talboys J., 1886, *India Under British Rule: From the Foundation of the East India Company*, London: Macmillan and Co.

Wild, Robert., and Christopher McLeod (ed.), 2008, *Sacred Natural Sites: Guidelines for Protected Area Managers*, Gland and Paris: International Union for Conservation of Nature and Natural Resources and UNESCO.

Wilford, Francis, 1805, 'An Essay on the Sacred Isles in the West, with other Essays connected with that Work', *Asiatick Researches*, 8: 245-75.

Wilhelm, Janine, 2016, *Environment and Pollution in Colonial India*, Oxon: Routledge.

SECONDARY SOURCES (HINDI)

Agarwal, Vasudev Sharan, 1956, Pāṇini kālīn Bhāratavarṣa (Bharata during Panini's Period), Varanasi: Chowkhamba Vidyabhavan.

Jayaswal, Vidula, 2011, *Adikashi Se Varanasi Tak (From Adi Kashi to Varanasi)*, New Delhi: Aryan Books International.

Chandra Moti, 1962 (fourth edition), *Kāśī kā Itihās (History of Kāśī)*. Varanasi: Vishwavidyalaya Prakashan.

Sukul, Kuber Nath, 1977, *Varanasi-Vaibhav* (*The Glory of Varanasi*), Patna: Bihar-Rashtrabhasha-Parishad.

Upadhyaya, Baldeva, 1978, *Purāṇavimarśa* (*Authentic Solution of the Critical Problems of Purāṇas*) (second edition). Varanasi: Chowkhamba Vidyabhawan.

Vishwakarma, Ishwar Sharan, 1987, *Kāśī kā Aitihāsika Bhūgōla* (*Historical Geography of Kāśī*), Delhi: Ramanand Vidya Bhawan.

SURVEYS AND REPORTS

Basic Texts of the 1972 World Heritage Convention, UNESCO World Heritage Centre, October 2019 (edition), France: UNESCO.

Centre for Science and Environment, *Churning Still Water*, 2012, New Delhi.

CRISIL Risk and Infrastructure Solutions Limited, 2015, City Development Plan for Varanasi, 2041, Final City Development Plan.

Draft National Water Framework Bill, 2016, Ministry of Jal Shakti Department of Water Resources, River Development and Ganga Rejuvenation, Government of India.

Fact Finding Committee on Mumbai Floods, Final Report, volume I, March 2006, Government of Maharashtra, India.

Guidelines for HRIDAY: Heritage City Development & Augmentation Yojana, 2015, Ministry of Urban Development. Government of India.

Guidelines for Jawaharlal Nehru National Urban Renewal Mission (JNNURM), December 2005, Ministry of Urban Development, New Delhi: Government of India.

Indian National Trust for the Art and Cultural Heritage (INTACH), 2016, Heritage City Development and Augmentation Yojana-HRIDAY. Preparation of City HRIDAY Plan including Civic Infrastructure Development Plan, New Delhi.

Joshi, Esha Basanti, 1965, Uttar Pradesh District Gazetteers, Varanasi, Lucknow: Government of Uttar Pradesh.

National Water Policy, 2012, Ministry of Jal Shakti Department of Water Resources, River Development and Ganga Rejuvenation, Government of India.

Operational Guidelines for HRIDAY: Heritage City Development & Augmentation Yojana, Ministry of Urban Development, Government of India, January 2015.

Press Release, India Meteorological Department, https://mausam.imd.gov.in (Accessed on 21 September 2020).

Report No. 15 of 2012-13—Performance Audit of Jawaharlal Nehru National Urban Renewal Mission, Ministry of Housing and Urban Poverty Alleviation. The Comptroller and Auditor General of India.

Smart Cities: Mission Statement and Guidelines, 2015, Ministry of Housing and Urban Affairs, Government of India.

Smart City Proposal Varanasi, 2016, Ministry of Urban Development, Government of India.

The Comptroller and Auditor General of India, Report No. 15 of 2012-13—Performance Audit of Jawaharlal Nehru National Urban Renewal Mission, Ministry of Housing and Urban poverty Alleviation, India.

The Uttar Pradesh Pond Development, Protection and Conservation Authority Bill, 2017 (Third Draft), Minor Irrigation Department, Government of Uttar Pradesh.

Town and Country Planning Department, Uttar Pradesh, India, http://uptownplanning.gov.in/post/en/introduction-of-development-area-varanasi, (Accessed on 17 September 2020).

World Population Prospects 2019: Highlights, United Nations, Department of Economic and Social Affairs. Population Division.

Urban Flooding: Standard Operating Procedure, 2017, Town and Country Planning Organization, Ministry of Urban Development, Government of India.

NEWSPAPER ARTICLES (ONLINE)

Chandran, Rahul, 3 June 2009, JNNURM Projects Lag Over Land, Utilities and Personnel Problem, https://www.livemint.com/Politics/AxEtlFuZihH8e89CG3sfeI/JNNURM-projects-lag-over-land-utilities-and-personnel-probl.html (Accessed on 21 October 2020).

Donthi, Narasimha Reddy, 30 October 2020, in *Hyderabad, an Ocean of Concrete is Turning Streets into Rivers After Heavy Rains*, Scroll, https://scroll.in/article/977145/an-ocean-of-concrete-is-washing-across-hyderabad-turning-its-streets-into-rivers-after-heavy-rains (Accessed on 1 November 2020).

Iyer, Harish, 31 August 2017, 'Mumbai Rains: 26 July 2005 to 29 August 2017: What Have We Learnt?' *The Quint* (online edition). https://www.thequint.com/news/india/mumbai-rains-2017-what-have-we-learnt-since-2005 (Accessed on 22 September 2020).

'Heavy Rains in Mumbai Revive Memories of 26 July 2005, Fury, 26 July 2019', *The Economic Times*, https://economictimes.indiatimes.com/news/politics-and-nation/heavy-rains-in-mumbai-revive-memories-of-july-26-2005-fury/articleshow/70400761.cms Accessed on 1 November 2020.

Pathak, Subhash, 1 October 2019, 'Failed Drainage System Adds to Patna's Plight as Many Areas Remain Flooded', https://www.hindustantimes.com/patna/failed-drainage-system-adds-to-patna-s-plight-as-many-areas-remain-flooded/story-lmDNdkVm8zmfRmFp5BAo8K.html (Accessed on 21 October 2020).

Singh, Mitashi., and Upmanyu Varnika, 11 October 2019, 'Urban Flooding: The Case of Drowning Cities and Rising Vulnerability, 'Retrieved from: ww.downtoearth.org.in/blog/natural-disasters/urban-flooding-the-case-of-drowning-cities-and-rising-vulnerability-67203 (Accessed on 19 September 2020).

Shah, Jagan, January 2020, 'Fixing India's Cities Needs Cutting the Gordian Knot: Failure, Inaction are Roadblocks to Leveraging Urbanisation for Good', https://www.firstpost.com/india/fixing-indias-cities-needs-cutting-the-gordian-knot-failure-inaction-are-roadblocks-to-leveraging-urbanisation-for-good-7857231.html (Accessed on 21 October 2020).

ONLINE WEBSITES

Archaeological Survey of India, Patna Circle, Patna, http://www.asipatnacircle.gov.in/uttarpradesh.html

Digital South Asia Library and the Digital Dictionaries of South Asia (University of Chicago and the Center for Research Libraries) https://dsal.uchicago.edu/dictionaries/

India Meteorological Department, https://mausam.imd.gov.in

Town and Country Planning Department, Uttar Pradesh, India, http://uptownplanning.gov.in/post/en/introduction-of-development-area-varanasi

THE WORLD BANK

https://www.worldbank.org/en/news/feature/2013/06/19/india-climate-change-impacts

Index